Keine original LEGO® Bauanleitungen. LEGO® hat dieses Buch weder gesponsert noch autorisiert. In diesem Buch enthaltene Bauanleitungen beruhen nicht auf einer Lizenz oder Billigung durch die Rechtsinhaber der LEGO®-Kennzeichen und LEGO®-Produkte. Die Benutzung der LEGO®-Marken und -Produkte erfolgt ausschließlich zur Illustration der in diesem Buch enthaltenen Bauideen. Sämtliche Rechte an Gestaltungen und Marken stehen allein dem jeweiligen Rechtsinhaber zu.

I0516217

Never too old to play with bricks...

Martin Ludwig & Frank Müller

WW2 Wehrmacht custom building instructions Volume 2

to be build out of LEGO® Bricks

www.tredition.de

© 2017 Martin Ludwig & Frank Müller
Umschlag, Illustration: Frank Müller

Verlag tredition GmbH, Hamburg

ISBN
Paperback: 978-3-7439-2229-7
Hardcover: 978-3-7439-2230-3
e-Book: 978-3-7439-2231-0

Printed in Germany

Das Werk, einschließlich seiner Teile, ist urheberrechtlich geschützt. Jede Verwertung ist ohne Zustimmung des Verlages und des Autors unzulässig. Dies gilt insbesondere für die elektronische oder sonstige Vervielfältigung, Übersetzung, Verbreitung und öffentliche Zugänglichmachung.

Keine original LEGO® Bauanleitungen.
LEGO® hat dieses Buch weder gesponsert noch autorisiert. In diesem Buch enthaltene Bauanleitungen beruhen nicht auf einer Lizenz oder Billigung durch die Rechtsinhaber der LEGO®-Kennzeichen und LEGO®-Produkte. Die Benutzung der LEGO®-Marken und -Produkte erfolgt ausschließlich zur Illustration der in diesem Buch enthaltenen Bau Ideen. Sämtliche Rechte an Gestaltungen und Marken stehen allein dem jeweiligen Rechtsinhaber zu.

Alle hier abgebildeten Modelle sind bei der DPMA registriert.

Ebenfalls im tredition Verlag erschienen:
Bauanleitung Tiger & Königstiger

Also published by tredition:
Building instruction Tiger & Kingtiger tank

ISBN
Paperback: 978-3-7323-1027-2
Hardcover: 978-3-7323-1028-9
e-Book: 978-3-7323-1029-6

Ebenfalls im tredition Verlag erschienen:
WW2 Wehrmacht custom building instructions

Also published by tredition:
WW2 Wehrmacht custom building instructions

ISBN
Paperback: 978-3-7323-4183-2
Hardcover: 978-3-7323-4184-9
e-Book: 978-3-7323-4185-6

Ebenfalls im tredition Verlag erschienen:
Miniscale Wehrmacht Vehicles

Also published by tredition:
Miniscale Wehrmacht Vehicles

ISBN
Paperback: 978-3-7323-7917-0
Hardcover: 978-3-7323-7918-7
e-Book: 978-3-7323-7919-4

Ebenfalls im tredition Verlag erschienen:
Vietnam War Vehicles

Also published by tredition:
Vietnam War Vehicles

ISBN
Paperback: 978-3-7345-9080-1
Hardcover: 978-3-7345-9081-8
e-Book: 978-3-7345-9082-5

Martin Ludwig und Frank Müller sind zwei begeisterte Hobby AFOLs (Adult Fan of LEGO®) die sich auf den Nachbau von Militär Fahrzeugen spezialisiert haben. Die Fahrzeuge sind passend für die Größe der Figuren ausgelegt.
Alles begann mit einer Idee. Nach erfolgreichem Verkauf einiger Fahrzeuge bei eBay® wurde die Produktpalette stetig erweitert. Je nach Größe dauert das Design eines Fahrzeuges einige Wochen. Hier achten wir zum größten Teil auf die Funktionalität und die Verfügbarkeit der Teile. Natürlich können wir alles noch besser und komplexer bauen, aber nicht jedem Kunden stehen unsere Möglichkeiten zur Verfügung. Nach der Veröffentlichung der ersten drei Baubücher liegt der Fokus dieser Bauanleitungen auf Vietnam Kriegs Fahrzeugen. In diesem Buch enthalten sind die Bauanleitungen für:
- Marder 2 Camouflage (Seite 8 - 29)
- Panzer 2 - 5cm (Seite 30 - 45)
- RSO 2cm Flak (Seite 46 - 62)
- SDKFZ 250 (Seite 63 - 77)
- Stug 3 (Seite 78 - X)

Martin Ludwig and Frank Müller are two hobby LEGO® builders from Germany. They spend their time building military models out of LEGO® bricks. It all started with an idea. After selling a few vehicles on eBay®, they started a successful business. The design of each vehicle takes a few weeks. Functionality and accurate size of vehicles are the main goals. After the successful release of their first three books, they have now finished their fourth instruction book focused on Vietnam War Vehicles. This book features:
- Marder 2 Camouflage (Page 8 - 29)
- Panzer 2 - 5cm (Page 30 - 45)
- RSO 2cm Flak (Page 46 - 62)
- SDKFZ 250 (Page 63 - 77)
- Stug 3 (Page 78 - X)

Marder 2 Camouflage

Zum Bau des Modells benötigen Sie ca. 391 LEGO® Bausteine.
Länge: ca. 15,9 cm Breite: ca. 9,4 cm Höhe: ca. 8,9 cm

Marder 2 Camouflage

Requires approx. 391 LEGO® bricks.
length: ca. 15.9 cm width: ca. 9.4 cm height: ca. 8.9 cm

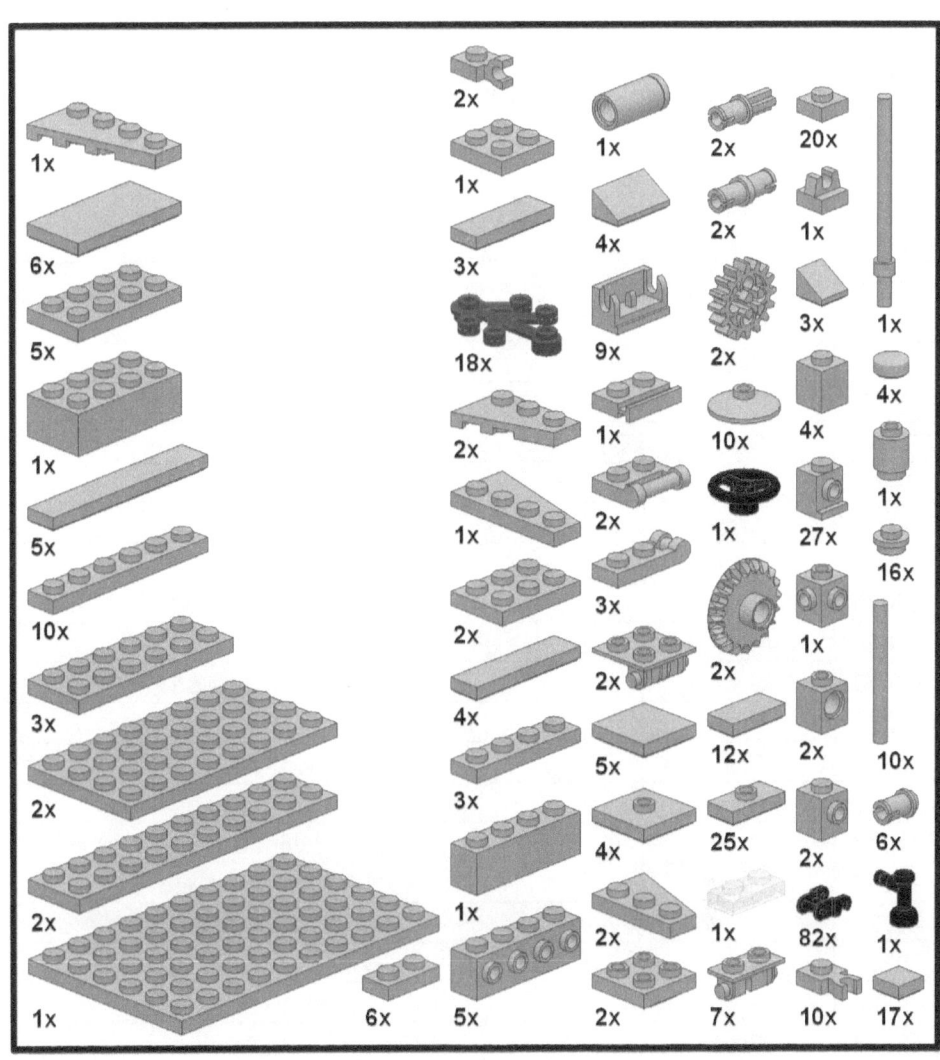

BLink ID	Color	Qty	Description
3035	Grey	2	Plate 4 x 8
3710	Grey	3	Plate 1 x 4
30414	Grey	5	Brick 1 x 4 with Studs on Side
4070	Grey	27	Brick 1 x 1 with Headlight
3033	Grey	1	Plate 6 x 10
3666	Grey	10	Plate 1 x 6
3937	Grey	9	Hinge 1 x 2 Base
3795	Grey	3	Plate 2 x 6
6636	Grey	5	Tile 1 x 6
3024	Grey	20	Plate 1 x 1
3023	Grey	6	Plate 1 x 2
3020	Grey	5	Plate 2 x 4
3068b	Grey	5	Tile 2 x 2 with Groove
87079	Grey	6	Tile 2 x 4 with Groove
3023	Transparent	1	Plate 1 x 2
60478	Grey	3	Plate 1 x 2 with Handle on End
6019	Grey	2	Plate 1 x 1 with Clip Horizontal
3069b	Grey	12	Tile 1 x 2 with Groove
3070b	Grey	17	Tile 1 x 1 with Groove
3794b	Grey	25	Plate 1 x 2 with Groove
98138	Grey	4	Tile 1 x 1 Round with Groove
43722	Grey	2	Wing 2 x 3 Right
43723	Grey	2	Wing 2 x 3 Left
3005	Grey	4	Brick 1 x 1
63864	Grey	3	Tile 1 x 3 with Groove
2423	Earth Green	18	Plant Leaves 4 x 3
3010	Grey	1	Brick 1 x 4
87087	Grey	2	Brick 1 x 1 with Stud on 1 Side
32028	Grey	1	Plate 1 x 2 with Door Rail
3832	Grey	2	Plate 2 x 10
4073	Grey	16	~Moved to 6141
4085c	Grey	10	Plate 1 x 1 with Clip Vertical

www.ww2custombrickmodels.de

BLink ID	Color	Qty	Description
30374	Grey	10	Bar 4L Light Sabre Blade
4740	Grey	10	Dish 2 x 2 Inverted
3001	Grey	1	Brick 2 x 4
2444	Grey	2	Plate 2 x 2 with Hole
6562	Grey	2	Technic Axle Pin
4019	Grey	2	Technic Gear 16 Tooth
4599	Black	1	Tap 1 x 1 with Hole in Spout
3022	Grey	1	Plate 2 x 2
3021	Grey	2	Plate 2 x 3
2431	Grey	4	Tile 1 x 4 with Groove
48336	Grey	2	Plate 1 x 2 with Handle Type 2
3938	Grey	7	Hinge 1 x 2 Top
85984	Grey	4	Slope Brick 31 1 x 2 x 0.667
6541	Grey	2	Technic Brick 1 x 1 with Hole
3673	Grey	2	Technic Pin
87407	Grey	2	Technic Gear 20 Tooth
2555	Grey	1	Tile 1 x 1 with Clip
54200	Grey	3	=Slope Brick 31 1 x 1 x 0.667
3062b	Grey	1	Brick 1 x 1 Round with Hollow Stud
4733	Grey	1	Brick 1 x 1 with Studs on Four Sides
30663	Black	1	Car Steering Wheel Large
4274	Grey	6	Technic Pin 1/2
63965	Grey	1	Bar 6L with Thick Stop
75535	Grey	1	Technic Pin Joiner Round
87580	Grey	4	Plate 2 x 2 with Groove
41770	Grey	1	Wing 2 x 4 Left
41769	Grey	1	Wing 2 x 4 Right
6134	Grey	2	Hinge 2 x 2 Top
3711	Black	82	Technic Chain Link

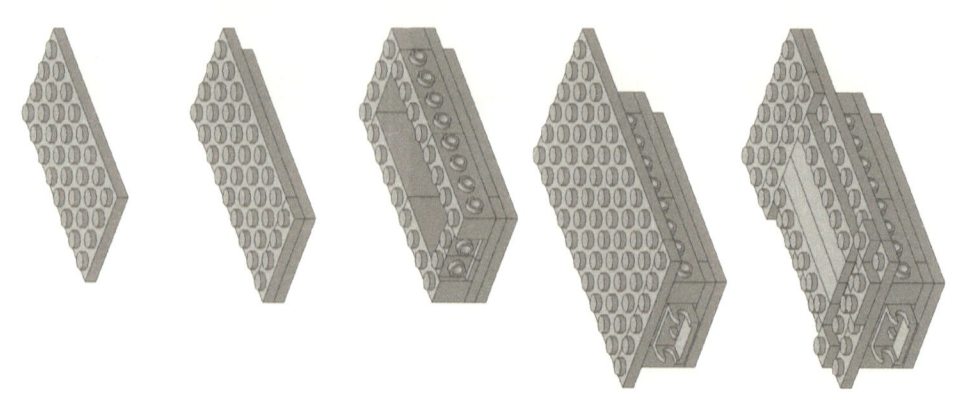
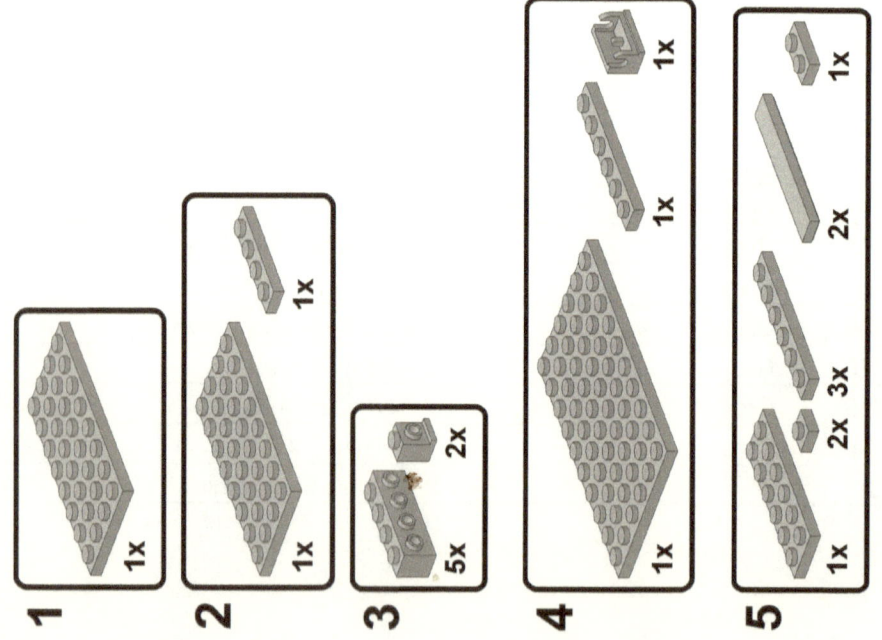

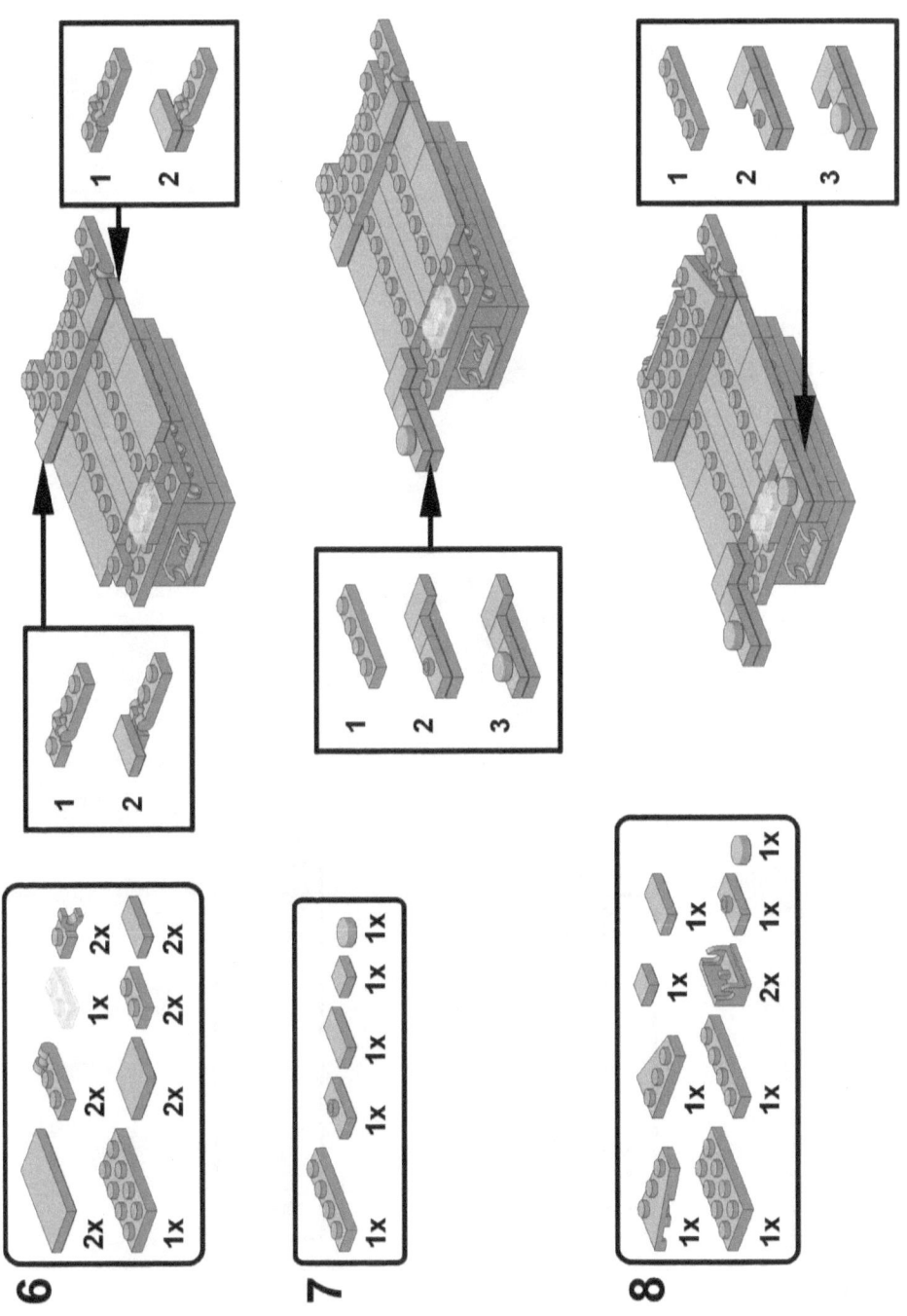

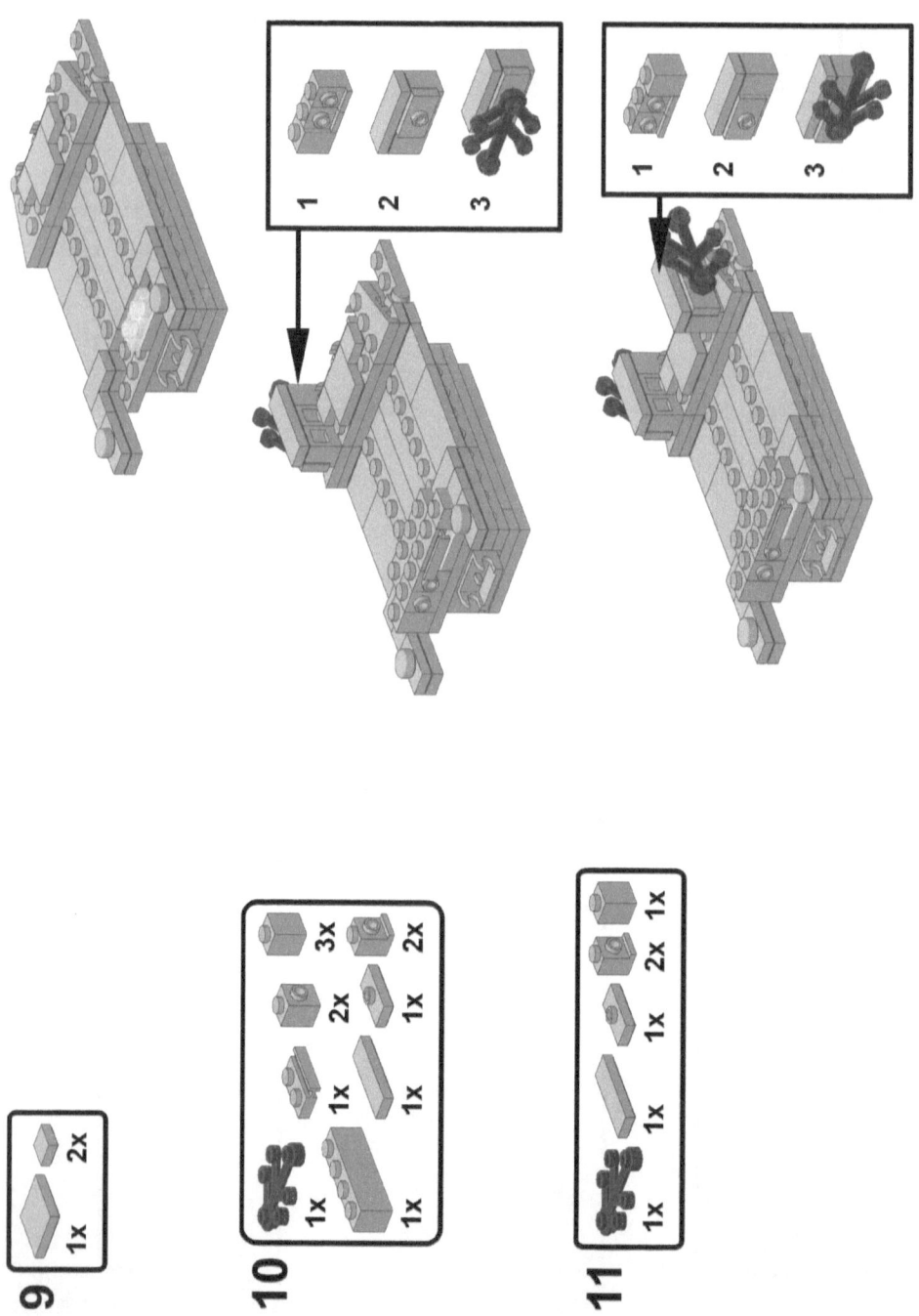

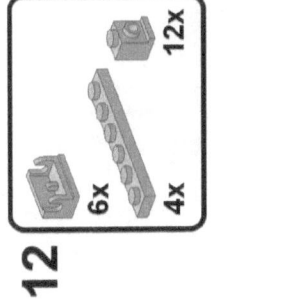
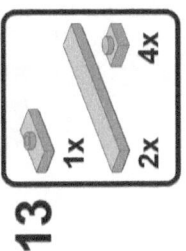

www.ww2custombrickmodels.de

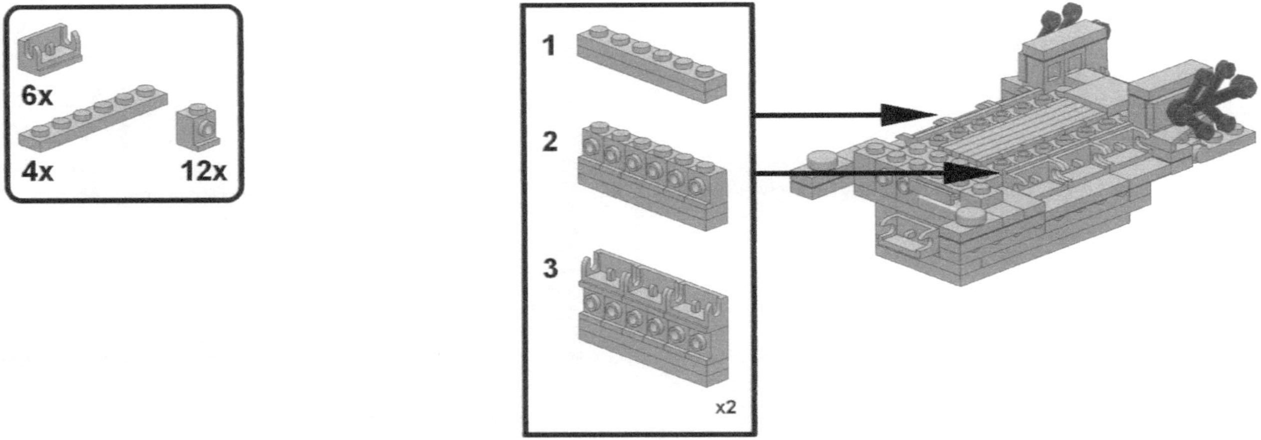

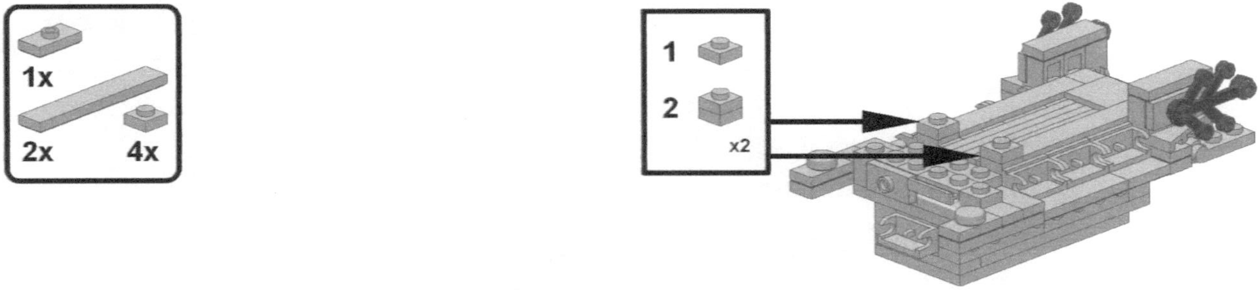

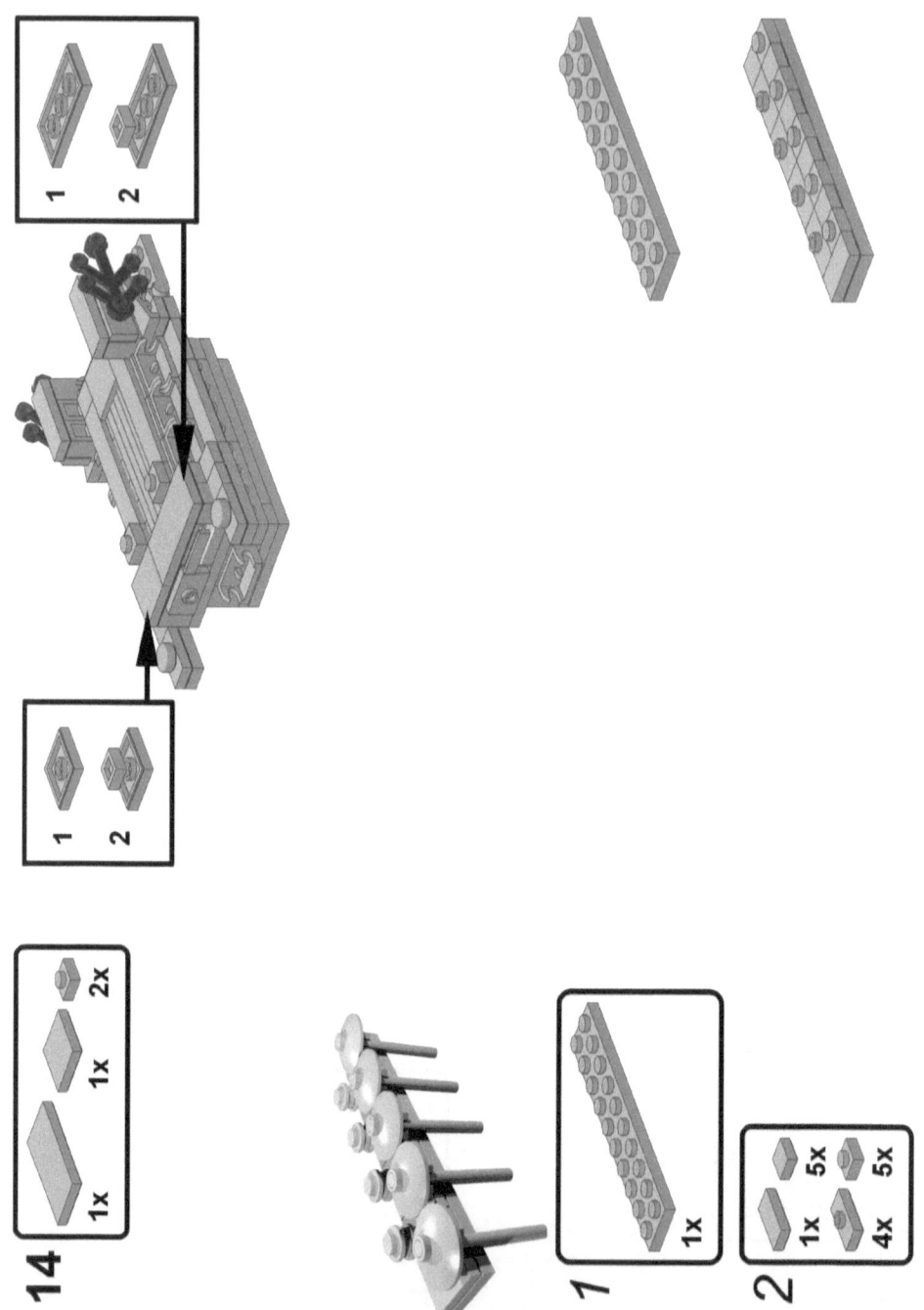

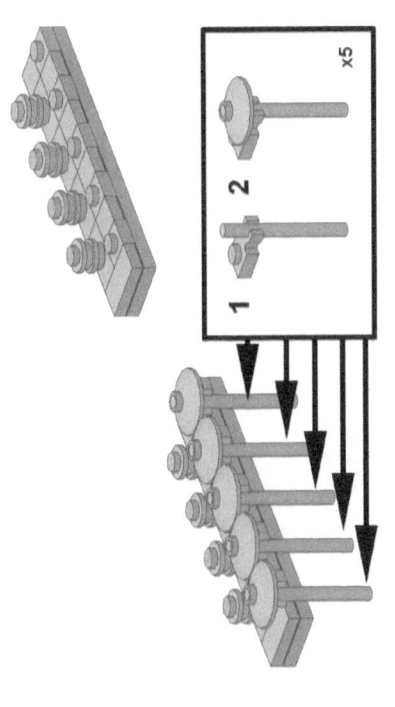
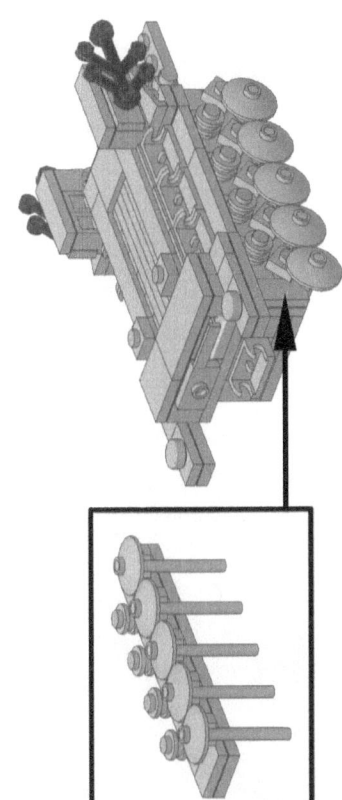

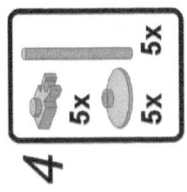

www.ww2custombrickmodels.de

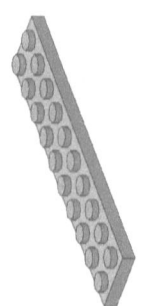
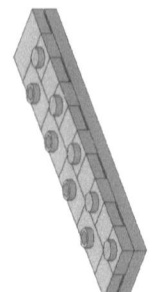
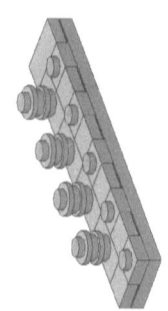

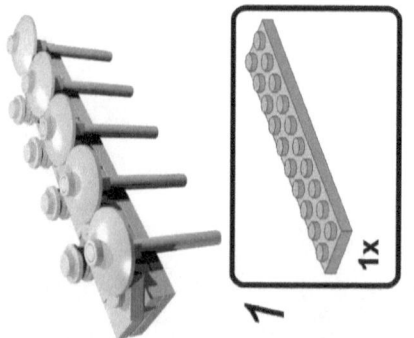
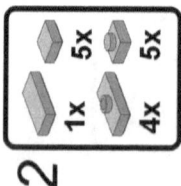

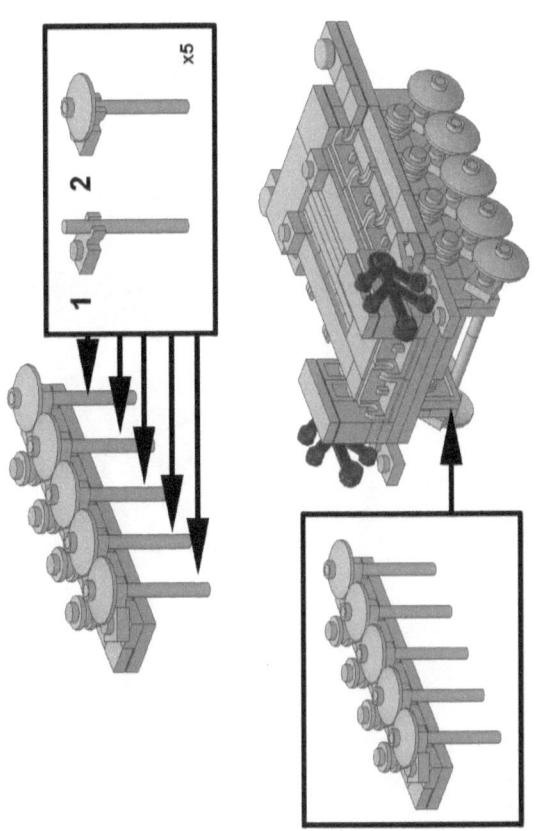
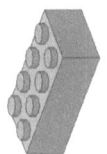
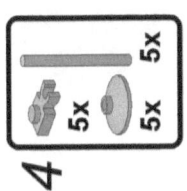

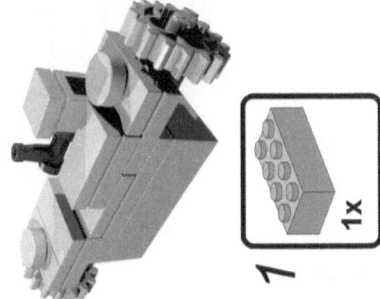

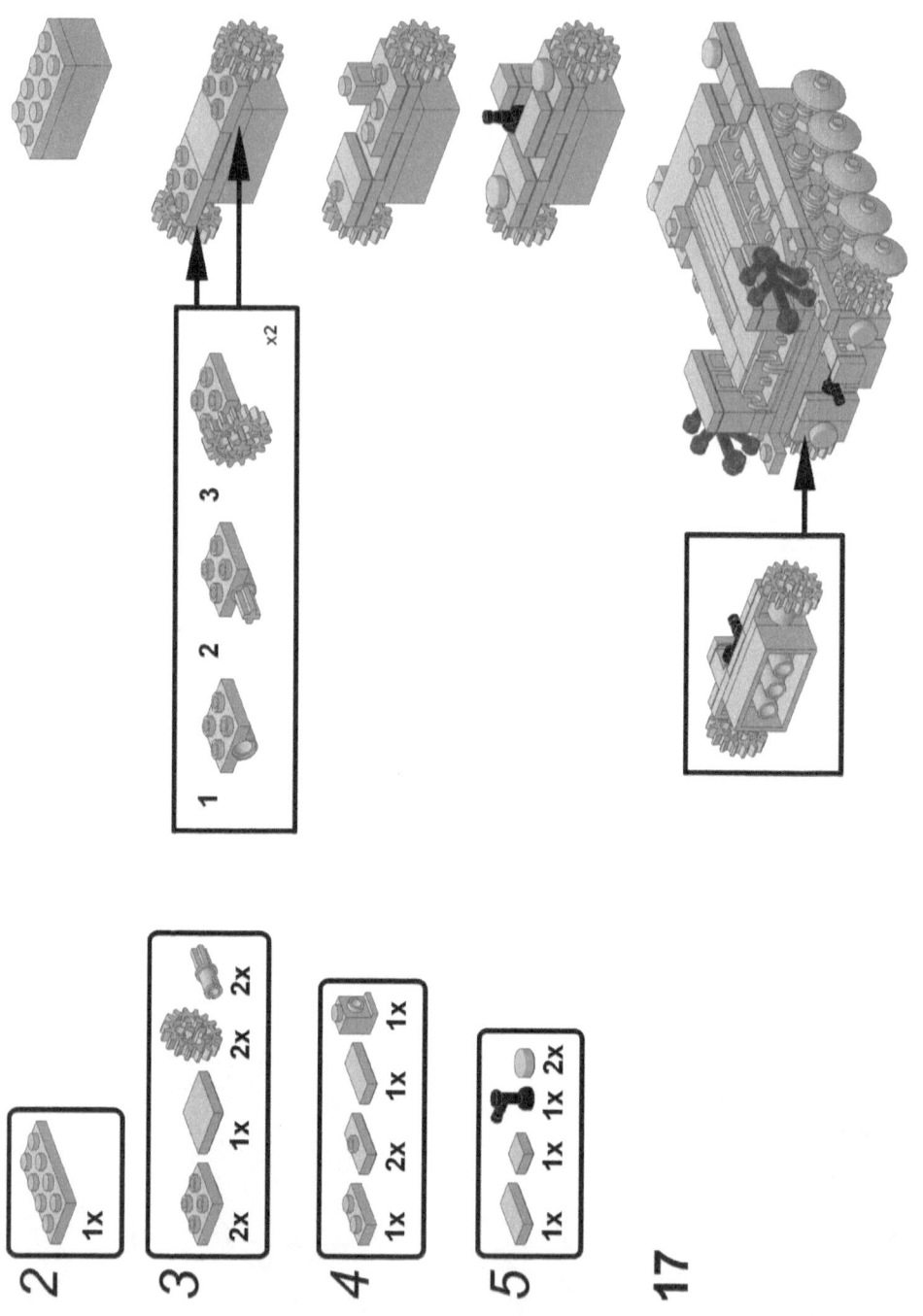

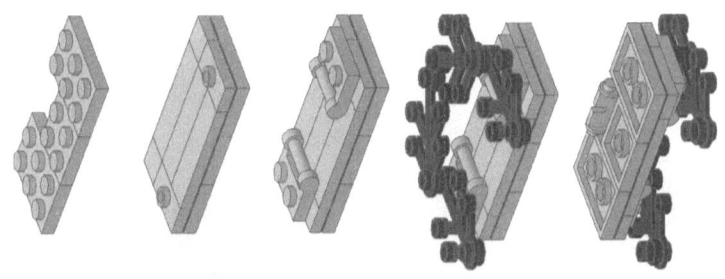
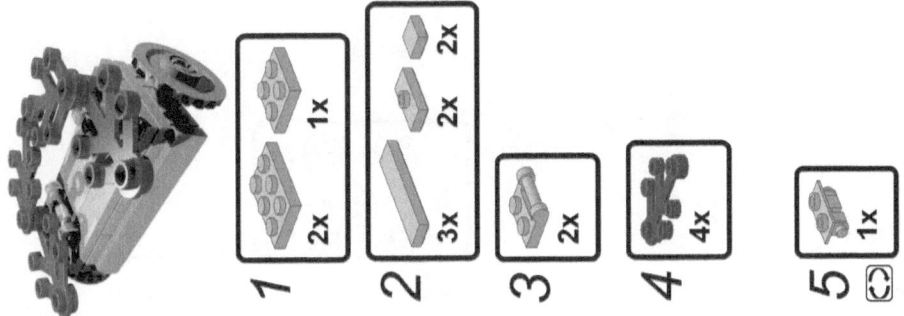

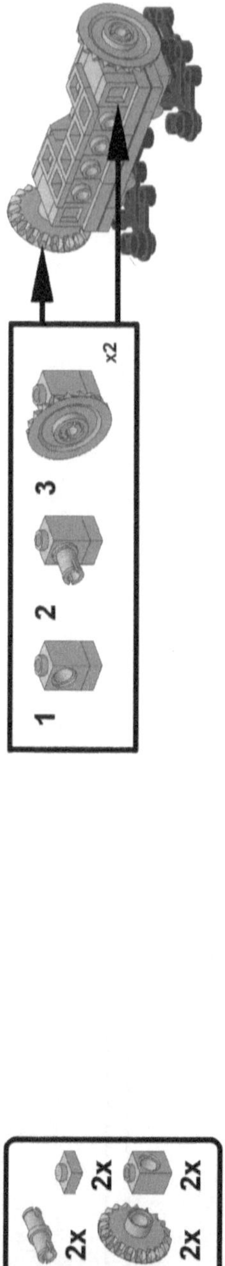

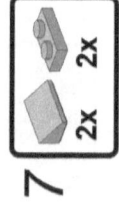
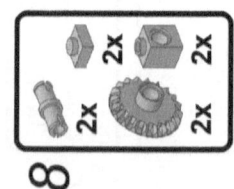

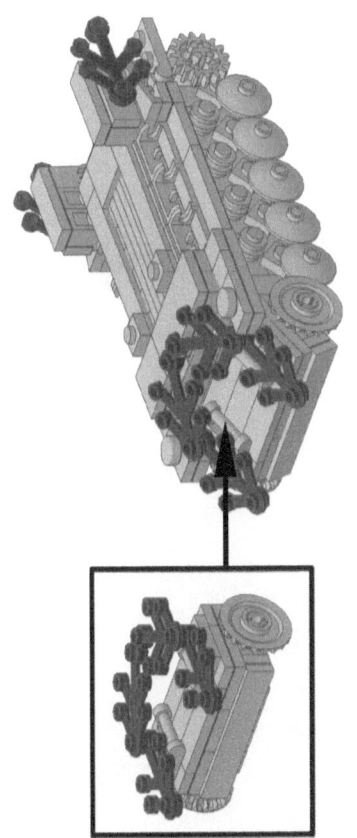

18

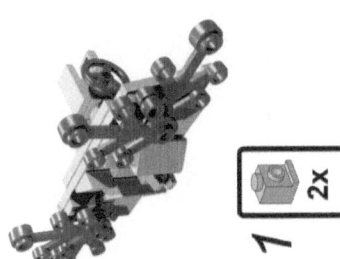
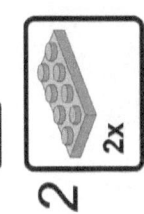

www.ww2custombrickmodels.de

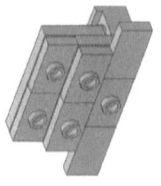 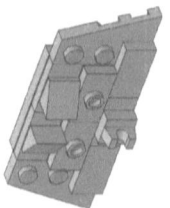 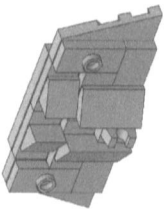 

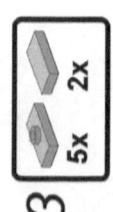
3: 2x, 5x

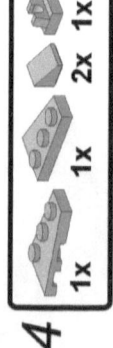
4: 1x, 2x, 1x, 1x

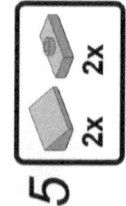
5: 2x, 2x

6: 4x

www.ww2custombrickmodels.de

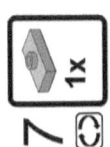

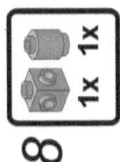

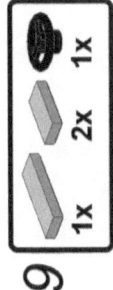

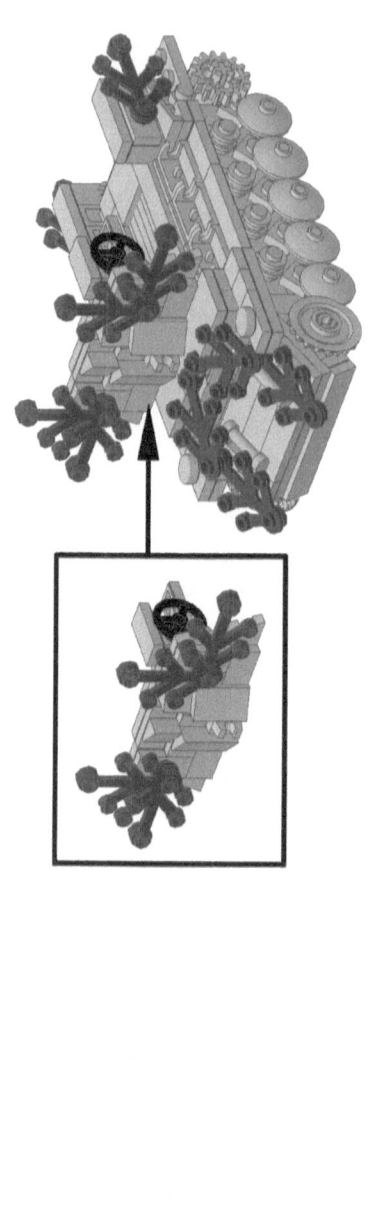

19

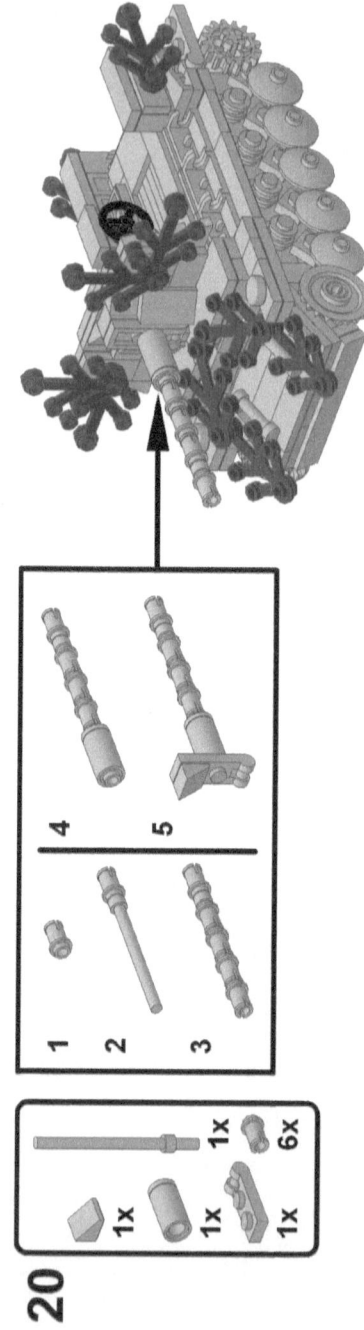

20

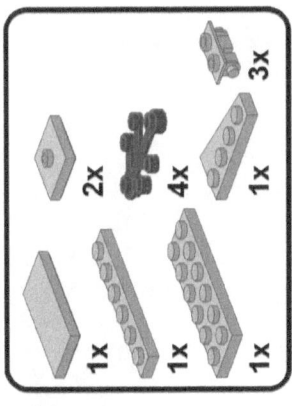

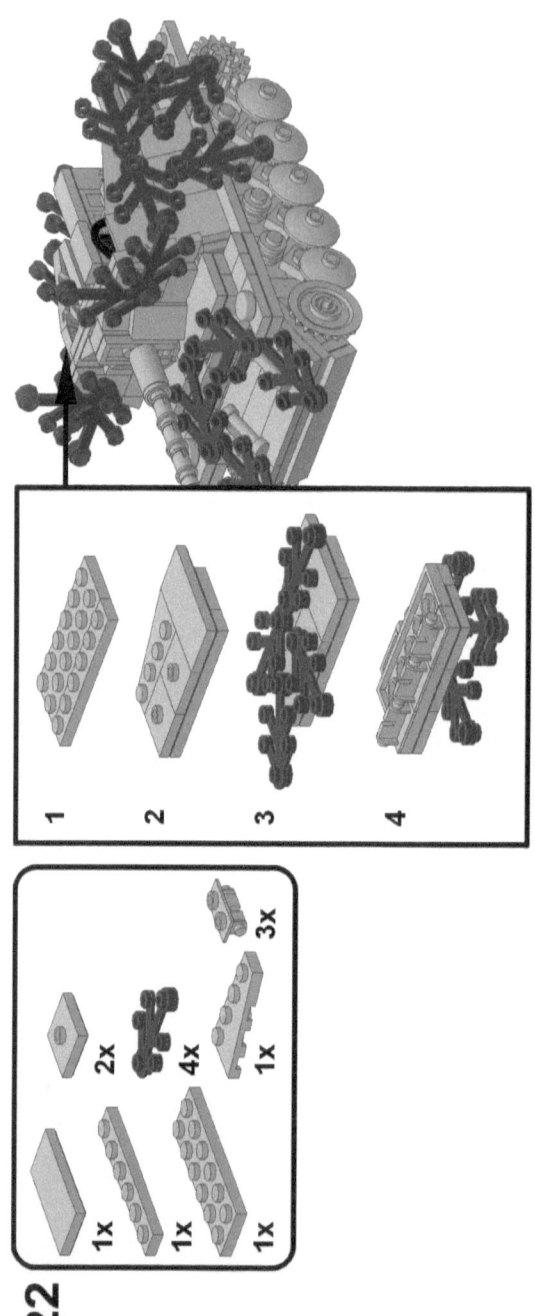

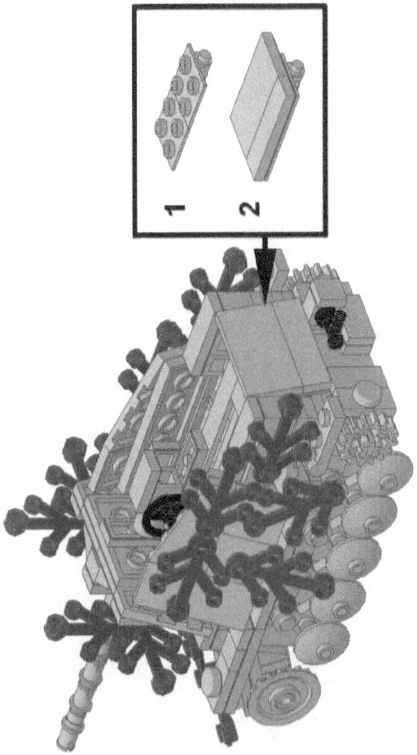

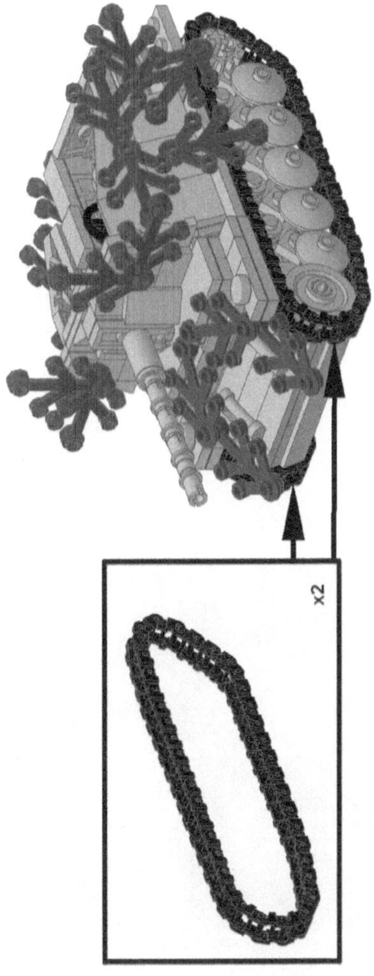

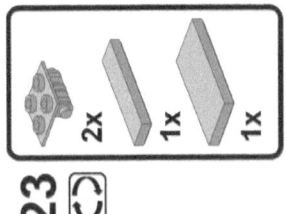

Panzer 2 - 5cm

Zum Bau des Modells benötigen Sie ca. 350 LEGO® Bausteine.
Länge: ca. 13,7 cm Breite: ca. 6,4 cm Höhe: ca. 6,7 cm

Panzer 2 - 5cm

Requires approx. 350 LEGO® bricks.
length: ca. 13.7 cm width: ca. 8.4 cm height: ca. 6.7 cm

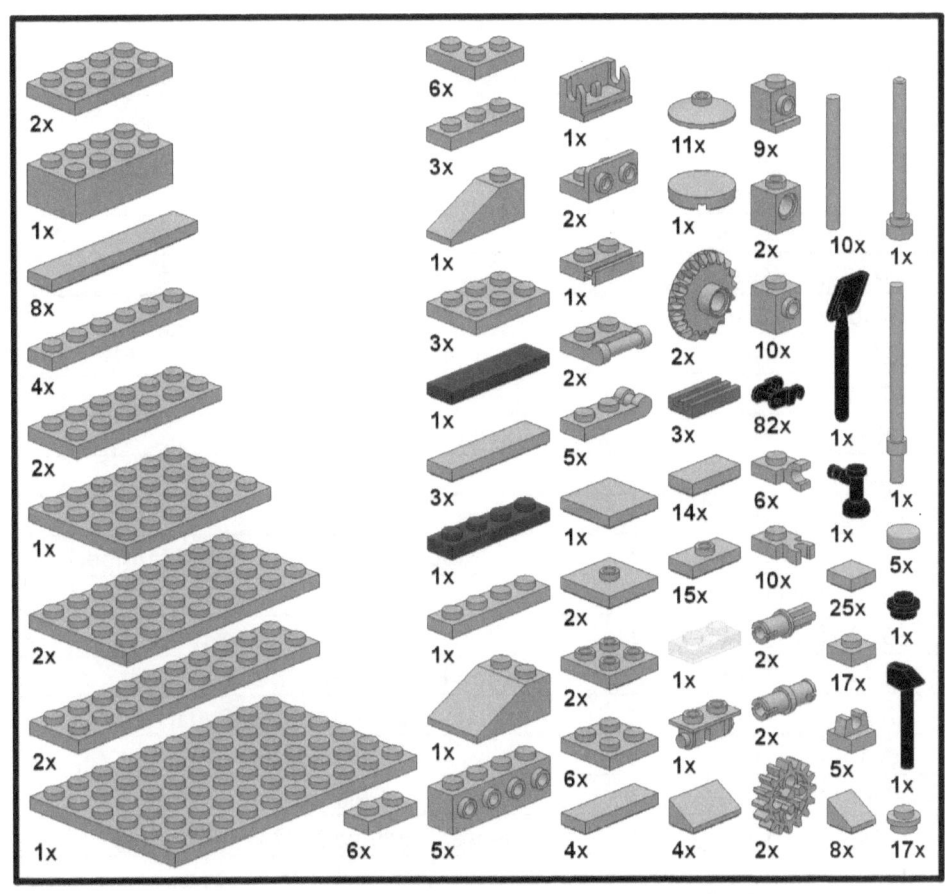

BLink ID	Color	Qty	Description
87087	Grey	10	Brick, Modified 1 x 1 with Stud on 1 Side
87580	Grey	2	Plate, Modified 2 x 2 with Groove and 1 Stud
6019	Grey	6	Plate, Modified 1 x 1 with Clip Horizontal
85984	Grey	4	Slope 30 1 x 2 x 2/3
3024	Grey	17	Plate 1 x 1
3022	Grey	6	Plate 2 x 2
3070b	Grey	25	Tile 1 x 1 with Groove
3023	Grey	6	Plate 1 x 2
54200	Grey	8	Slope 30 1 x 1 x 2/3
3623	Grey	3	Plate 1 x 3
63864	Grey	4	Tile 1 x 3
3021	Grey	3	Plate 2 x 3
3794	Grey	15	Plate, Modified 1 x 2 with 1 Stud (Jumper)
3957b	Grey	1	Antenna 1 x 4 - Flat Top
99780	Grey	2	Bracket 1 x 2 - 1 x 2 Inverted
30374	Grey	10	Bar 4L (Lightsaber Blade / Wand)
4740	Grey	11	Dish 2 x 2 Inverted (Radar)
3001	Grey	1	Brick 2 x 4
3020	Grey	2	Plate 2 x 4
2444	Grey	2	Plate, Modified 2 x 2 with Pin Hole
3069b	Grey	14	Tile 1 x 2 with Groove
3068b	Grey	1	Tile 2 x 2 with Groove
4070	Grey	9	Brick, Modified 1 x 1 with Headlight
4073	Black	1	Plate, Round 1 x 1 Straight Side
4599b	Black	1	Tap 1 x 1 without Hole in End
3749	Grey	2	Technic, Axle Pin without Friction
4019	Grey	2	Technic, Gear 16 Tooth
98138	Grey	5	Tile, Round 1 x 1
48336	Grey	2	Plate, Modified 1 x 2 with Handle on Side
14769	Grey	1	Tile, Round 2 x 2 with Bottom Stud Holder
3832	Grey	2	Plate 2 x 10
4073	Grey	17	Plate, Round 1 x 1 Straight Side

BLink ID	Color	Qty	Description
4085c	Grey	10	Plate, Modified 1 x 1 with Clip Vertical
6636	Grey	8	Tile 1 x 6
6541	Grey	2	Technic, Brick 1 x 1 with Hole
2555	Grey	5	Tile, Modified 1 x 1 with Clip
2431	Grey	3	Tile 1 x 4
3938	Grey	1	Hinge Brick 1 x 2 Top Plate Thin
60478	Grey	5	Plate, Modified 1 x 2 with Handle on End
3673	Grey	2	Technic, Pin without Friction Ridges Lengthwise
87407	Grey	2	Technic, Gear 20 Tooth Bevel with Pin Hole
3711	Black	82	Technic, Link Chain
3035	Grey	2	Plate 4 x 8
3710	Grey	1	Plate 1 x 4
3937	Grey	1	Hinge Brick 1 x 2 Base
3666	Grey	4	Plate 1 x 6
3023	Transparent	1	Plate 1 x 2
30414	Grey	5	Brick, Modified 1 x 4 with 4 Studs on 1 Side
3033	Grey	1	Plate 6 x 10
3795	Grey	2	Plate 2 x 6
2412b	Dark Grey	3	Tile, Modified 1 x 2 Grille
2420	Grey	6	Plate 2 x 2 Corner
3298	Grey	1	Slope 33 3 x 2
4286	Grey	1	Slope 33 3 x 1
32028	Grey	1	Plate, Modified 1 x 2 with Door Rail
3032	Grey	1	Plate 4 x 6
4522	Black	1	Minifig, Utensil Tool Mallet / Hammer
63965	Grey	1	Bar 6L with Stop Ring
3837	Black	1	Minifig, Utensil Shovel (Round Stem End)
3710	Brown	1	Plate 1 x 4
2431	Brown	1	Tile 1 x 4

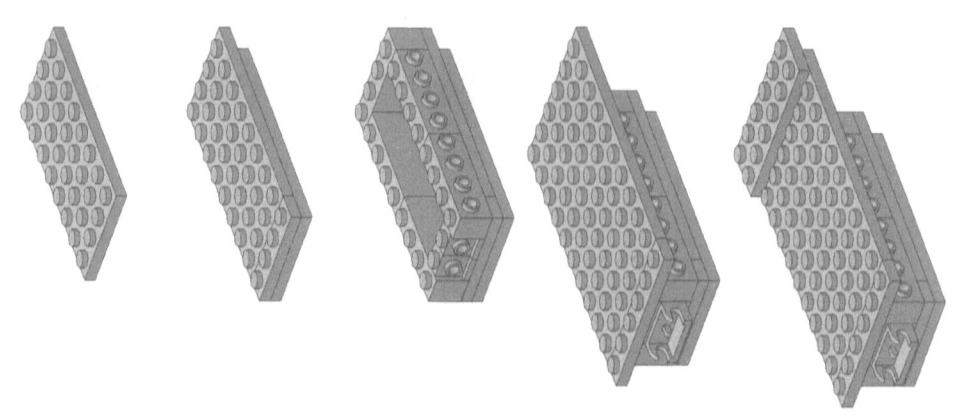

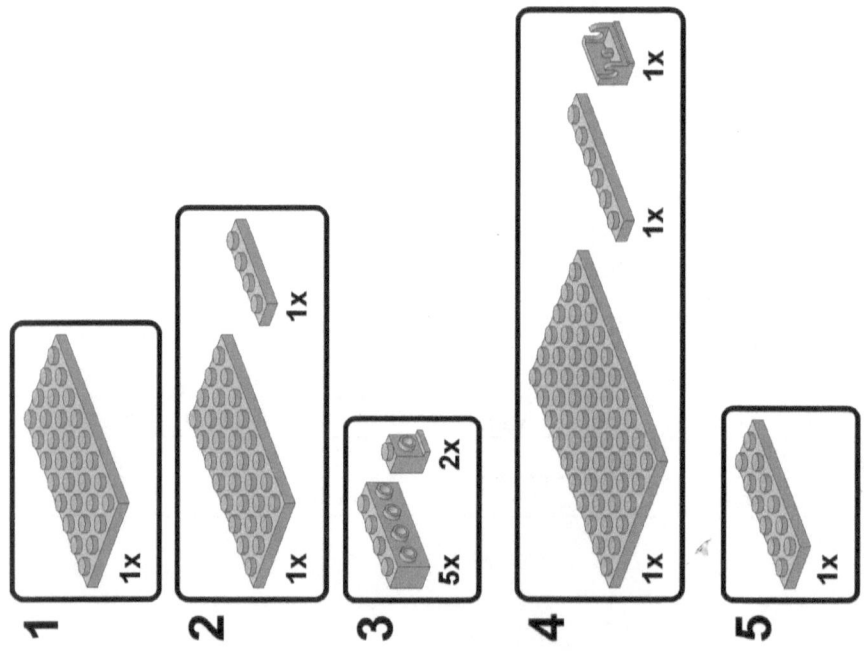

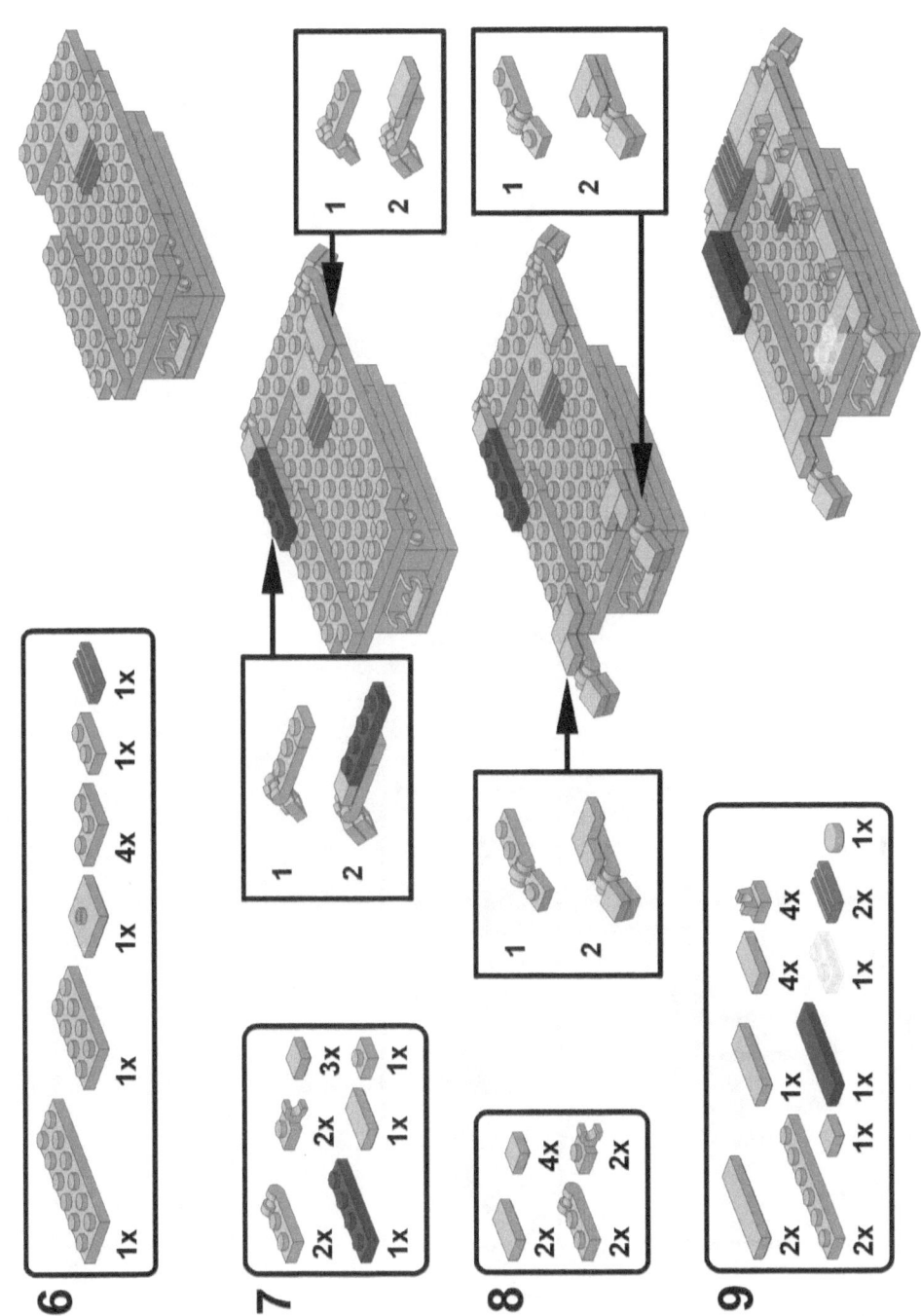

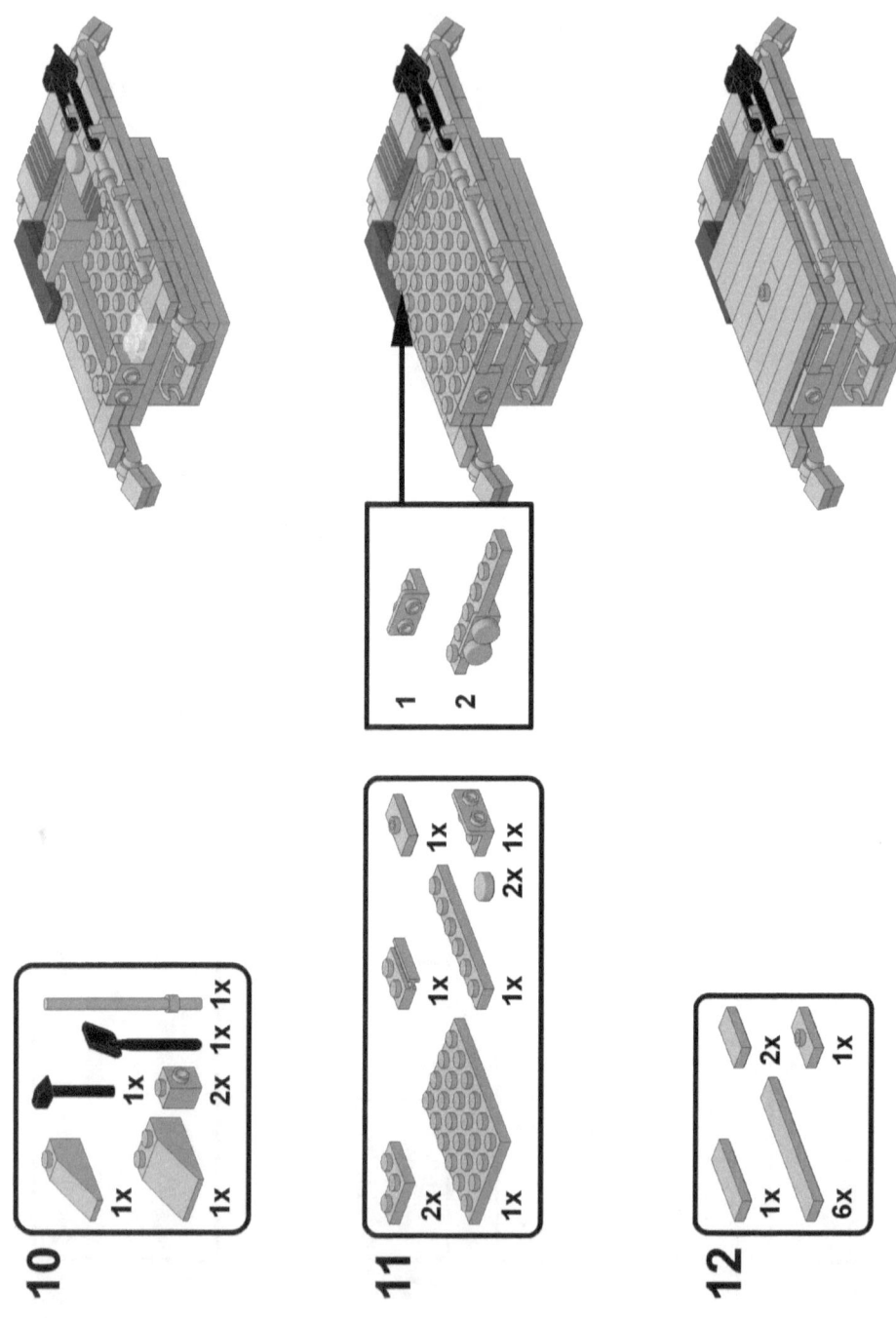

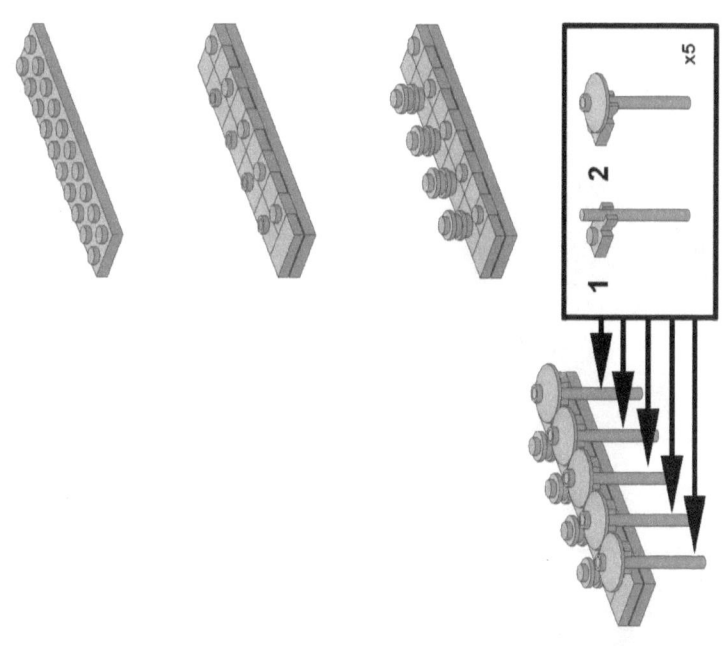
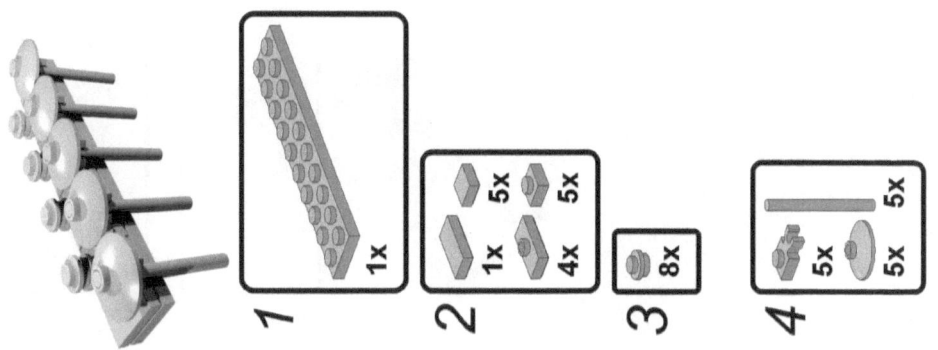

13

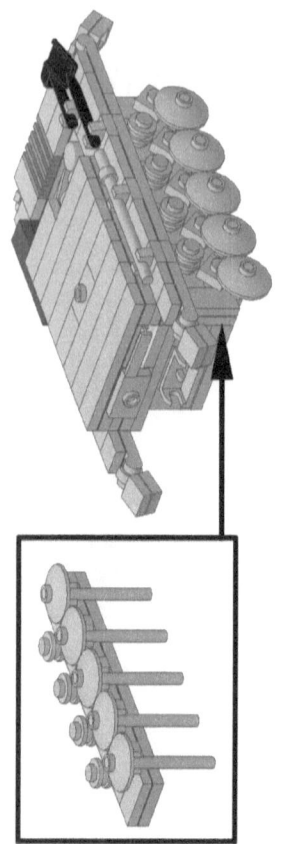
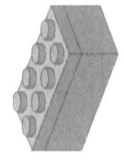

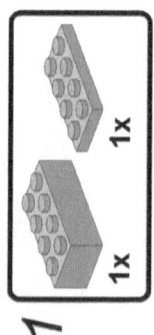
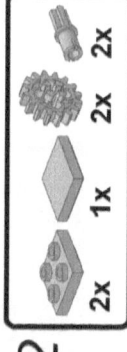

www.ww2custombrickmodels.de

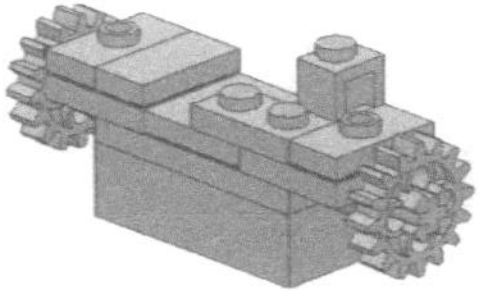

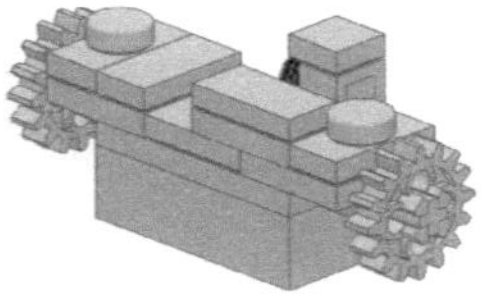

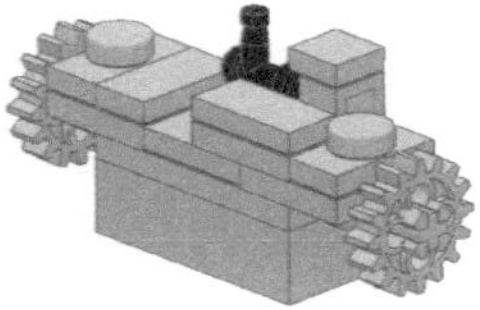

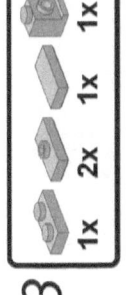
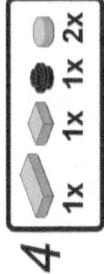

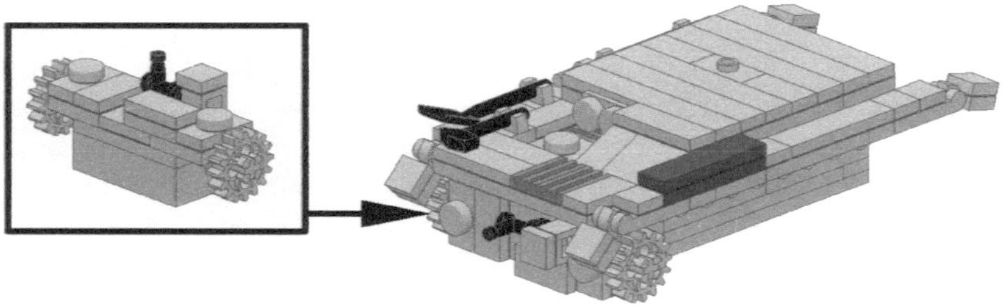

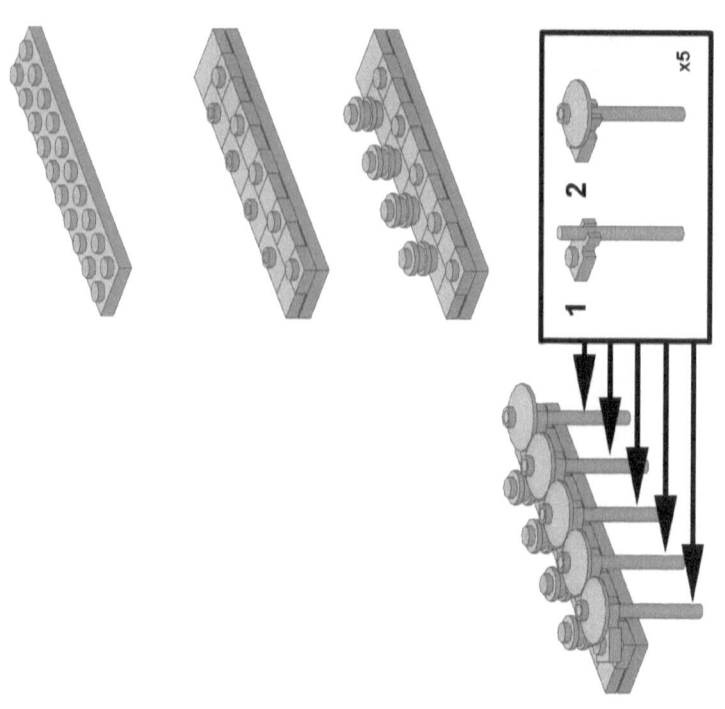
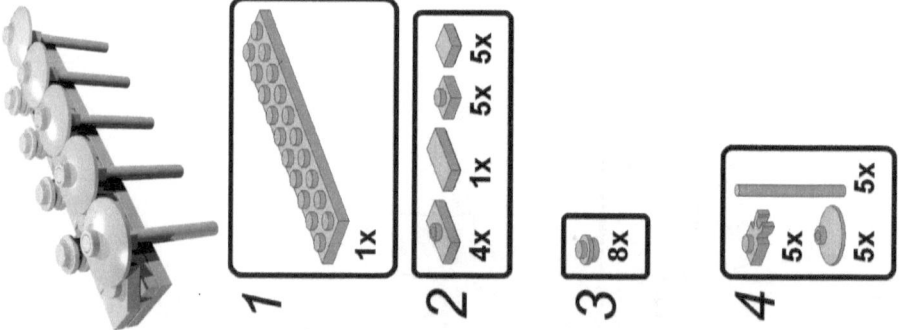

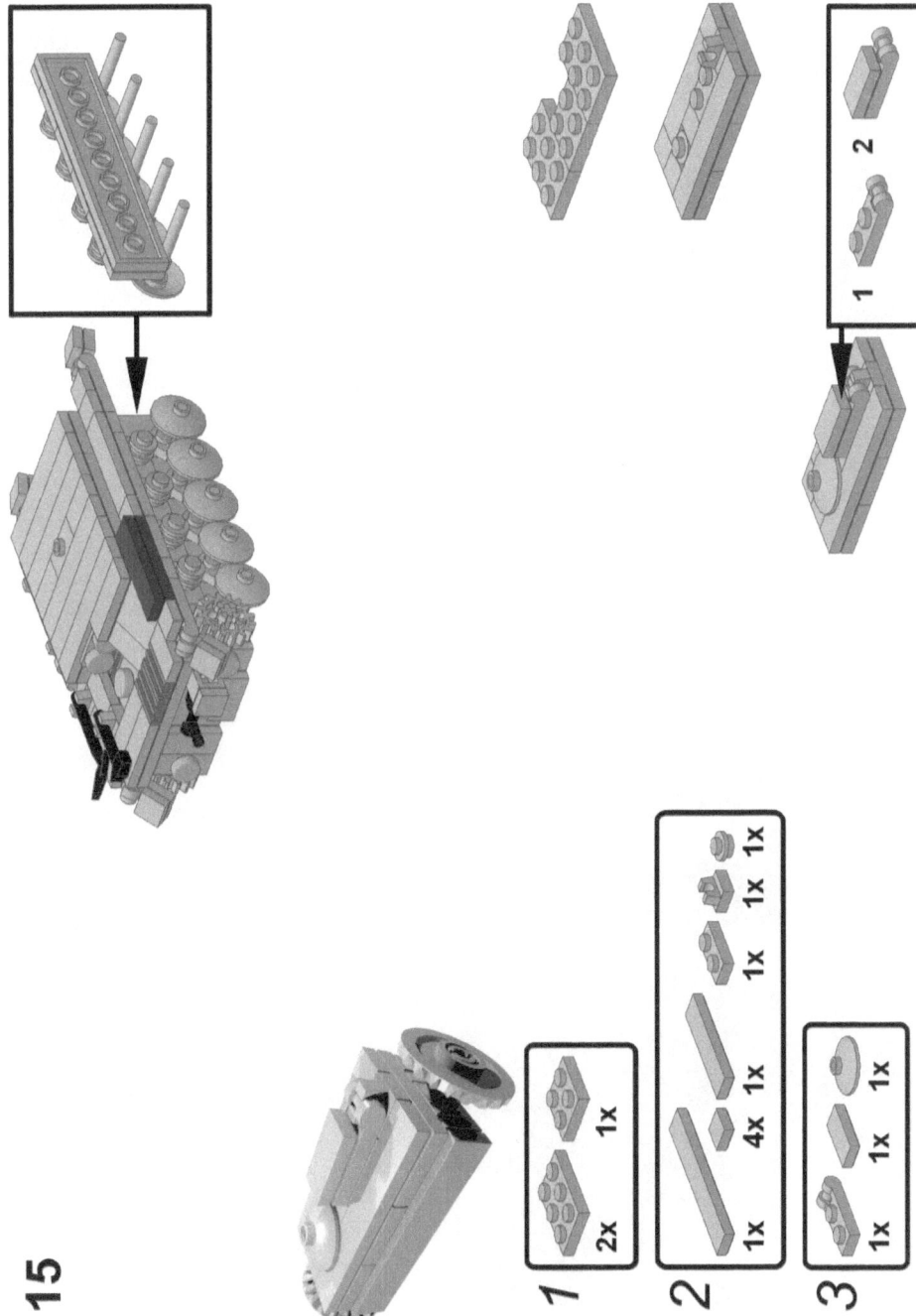

4
1x
5
6x
6
2x
2x
1
2
x2
7
2x
2x
2x
2x
1
2
3
x2
8
1x

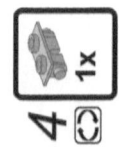

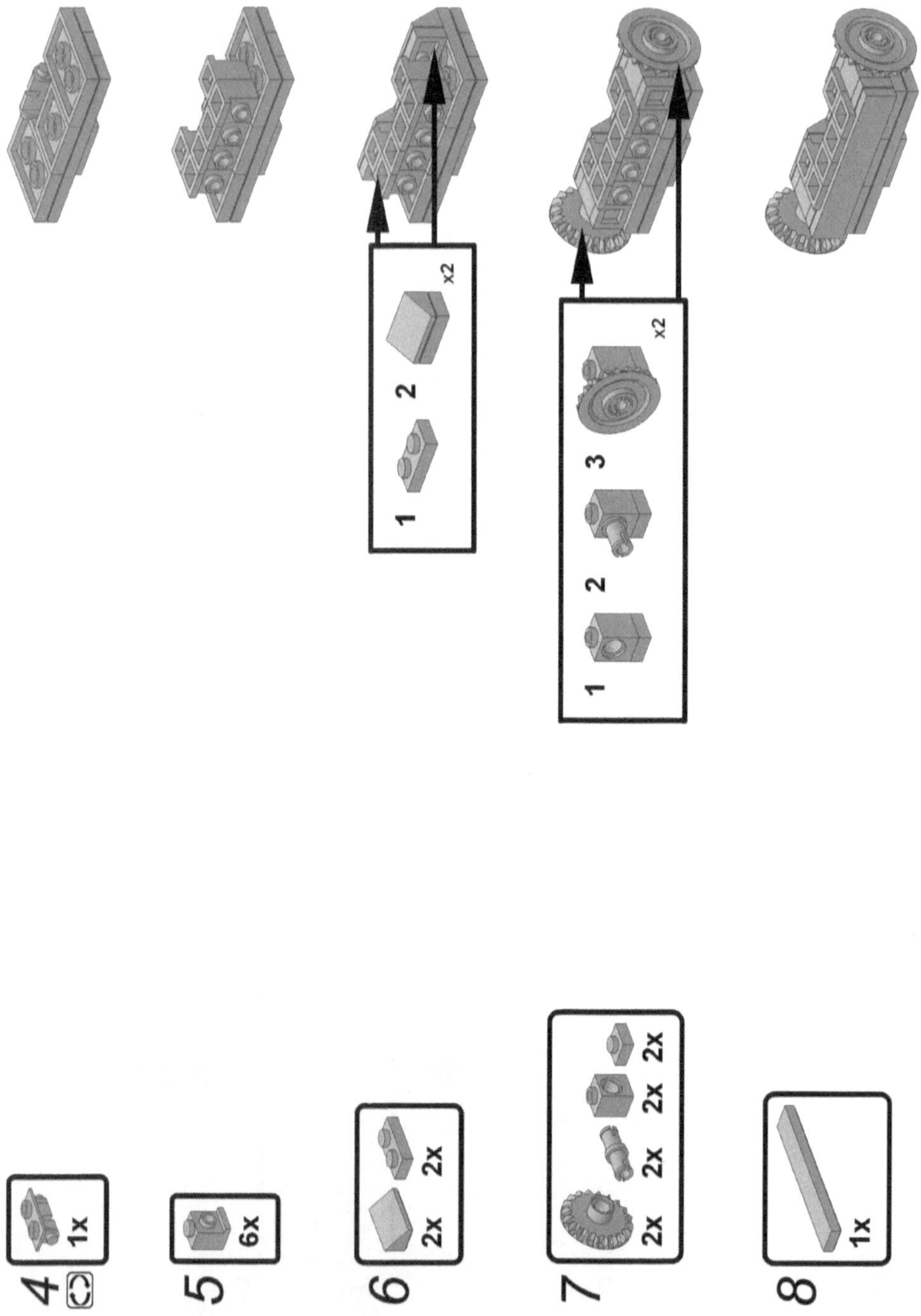

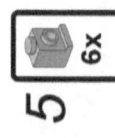

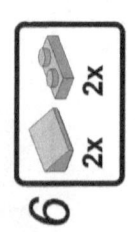

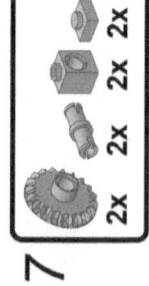

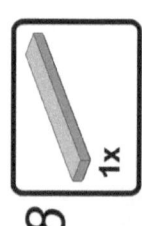

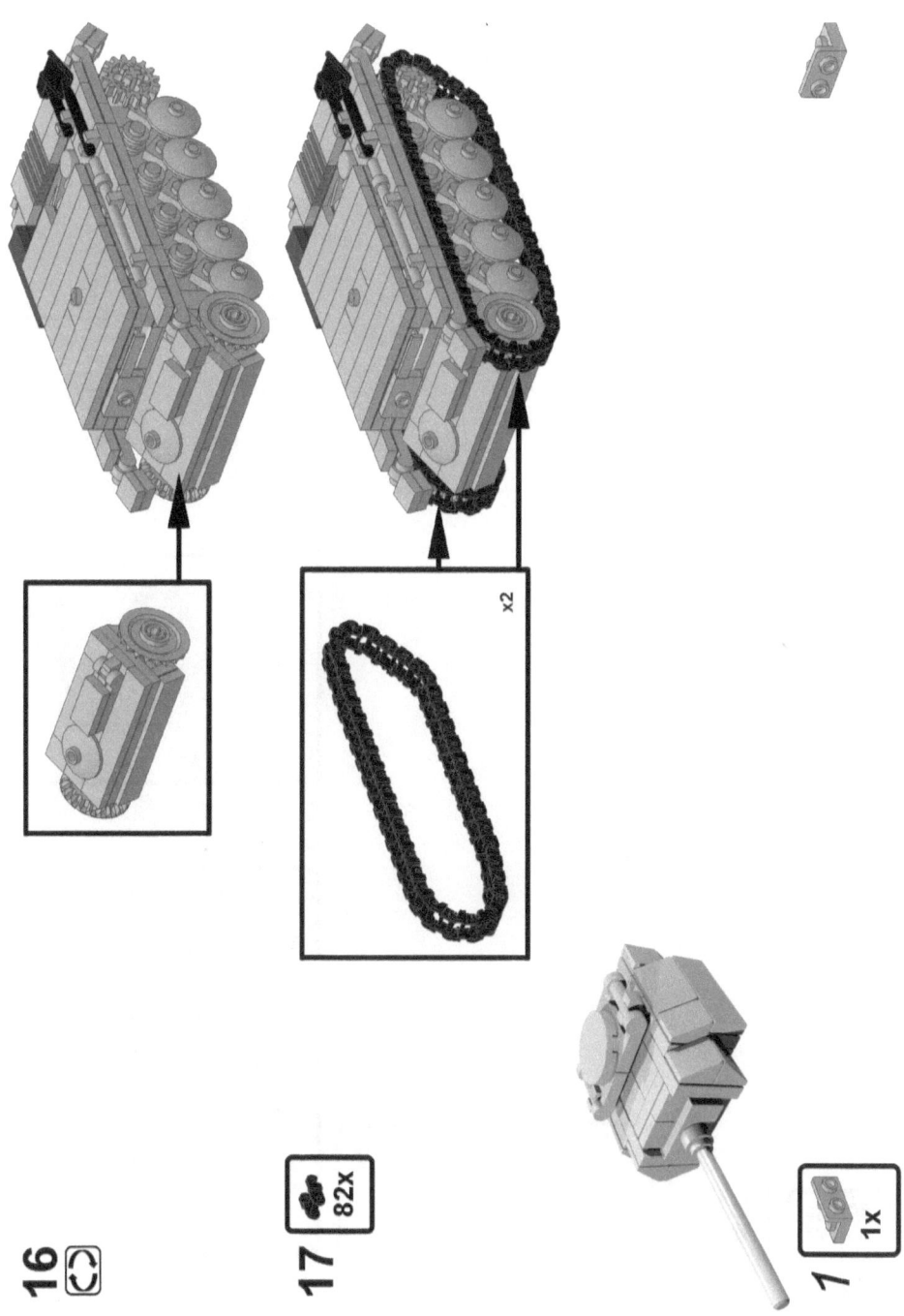

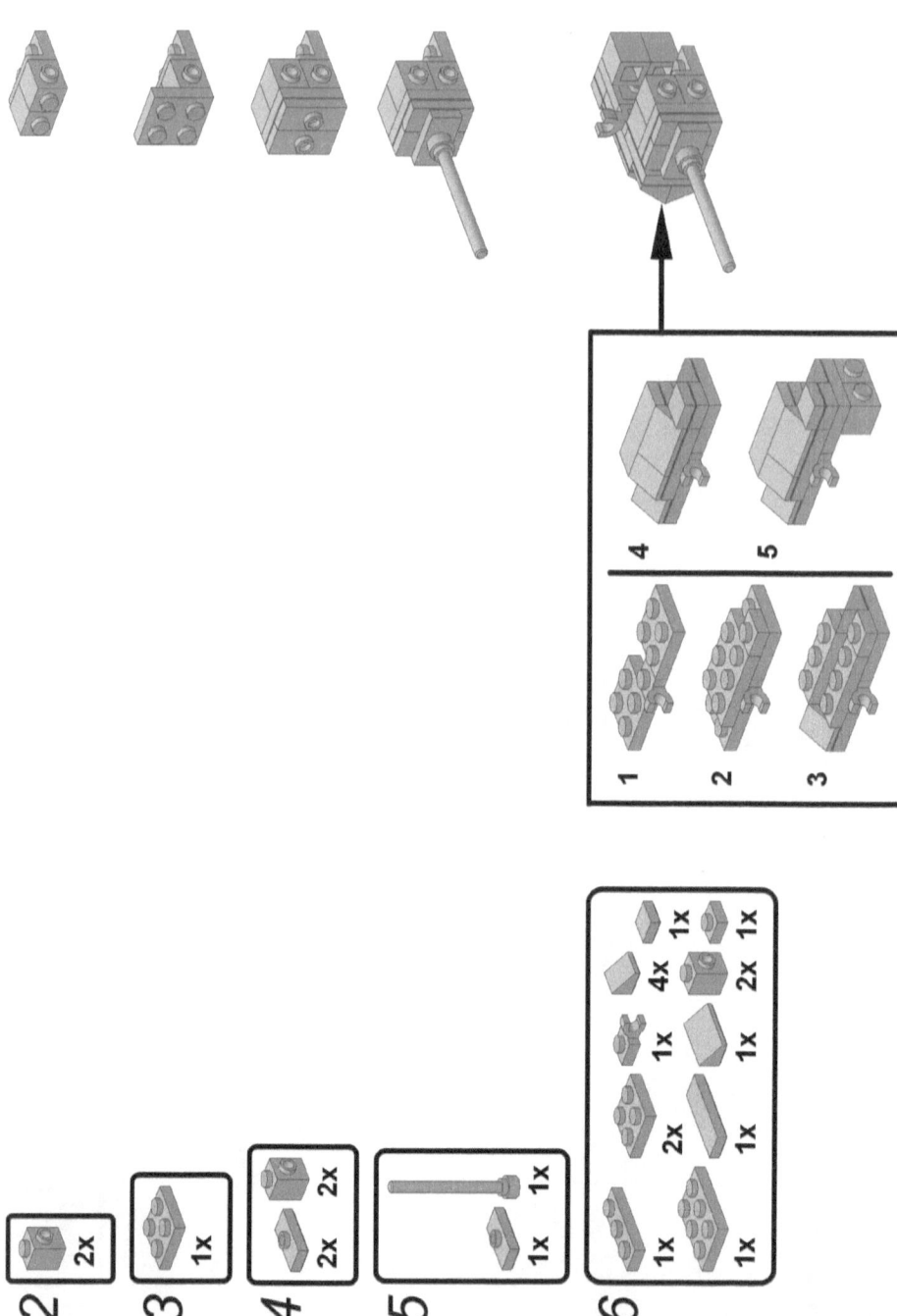

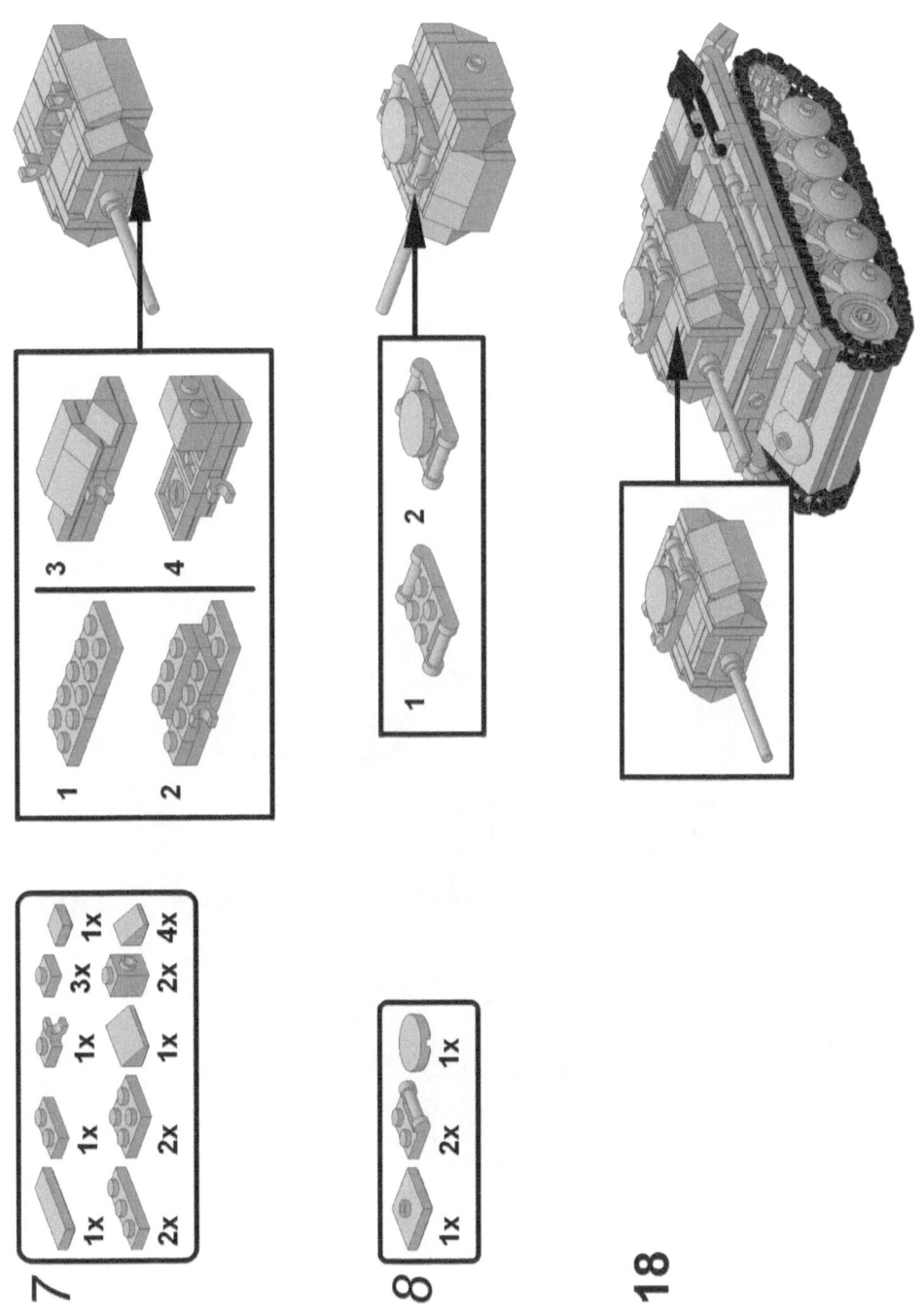

RSO 2cm Flak

Zum Bau des Modells benötigen Sie ca. 336 LEGO® Bausteine.
Länge: ca. 17,3 cm Breite: ca. 8,9 cm Höhe: ca. 8,3 cm

RSO 2cm Flak

Requires approx. 336 LEGO® bricks.
length: ca. 17.3 cm width: ca. 8.9 cm height: ca. 8.3 cm

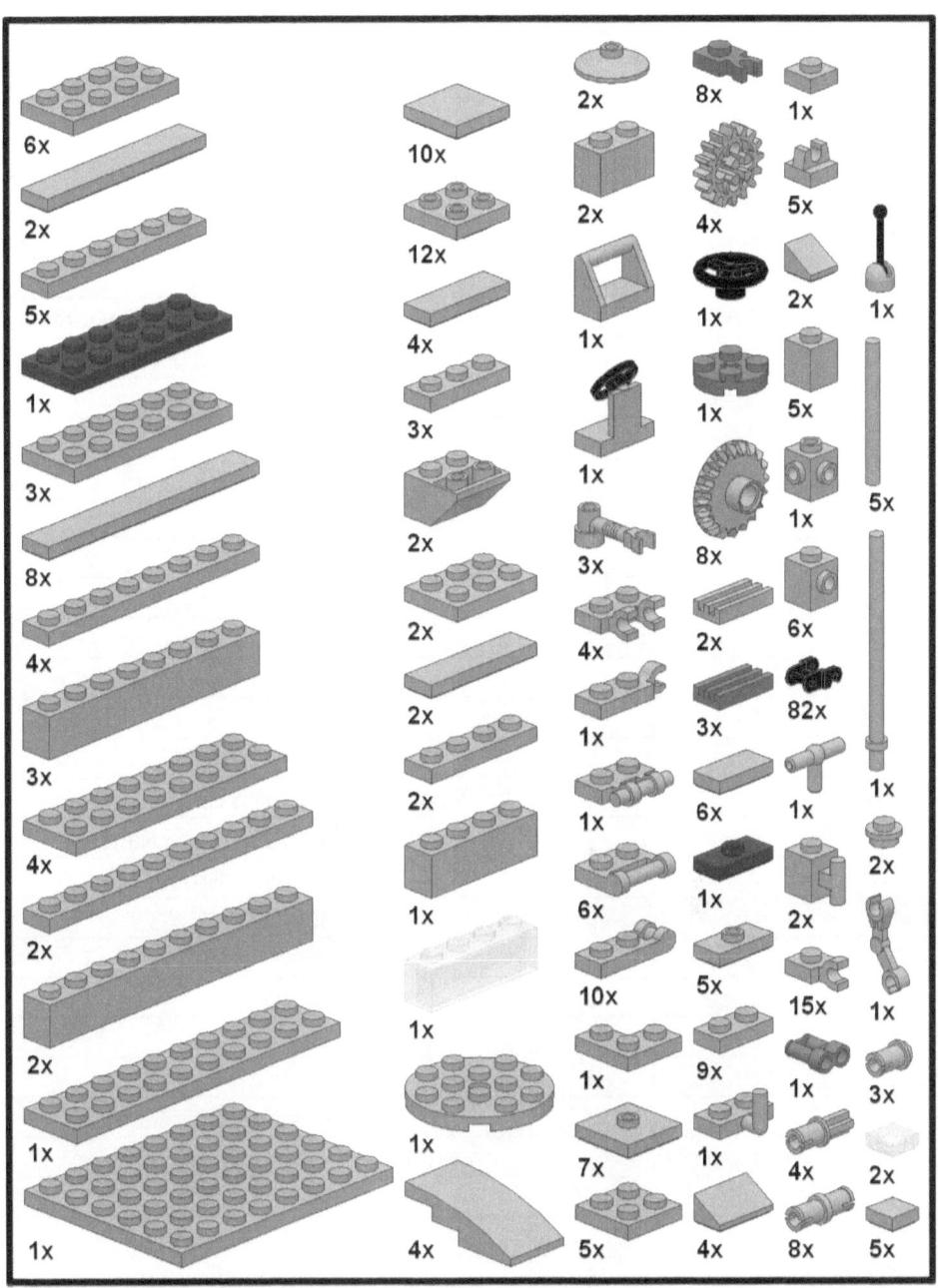

BLink ID	Color	Qty	Description
3829c01	Black	1	Vehicle, Steering Stand 1 x 2
3020	Grey	6	Plate 2 x 4
2444	Grey	12	Plate, Modified 2 x 2 with Pin Hole
3005	Grey	5	Brick 1 x 1
3460	Grey	4	Plate 1 x 8
3795	Grey	3	Plate 2 x 6
6019	Grey	15	Plate, Modified 1 x 1 with Clip Horizontal
3660	Grey	2	Slope, Inverted 45 2 x 2
3022	Grey	5	Plate 2 x 2
3832	Grey	1	Plate 2 x 10
6111	Grey	2	Brick 1 x 10
3623	Grey	3	Plate 1 x 3
3036	Grey	1	Plate 6 x 8
85984	Grey	4	Slope 30 1 x 2 x 2/3
3795	Brown	1	Plate 2 x 6
2412b	Grey	2	Tile, Modified 1 x 2 Grille
3666	Grey	5	Plate 1 x 6
3010	Grey	1	Brick 1 x 4
3069b	Grey	6	Tile 1 x 2 with Groove
2921	Grey	2	Brick, Modified 1 x 1 with Handle
3070b	Grey	5	Tile 1 x 1 with Groove
3710	Grey	2	Plate 1 x 4
2431	Grey	2	Tile 1 x 4
3008	Grey	3	Brick 1 x 8
3004	Grey	2	Brick 1 x 2
3034	Grey	4	Plate 2 x 8
93606	Grey	4	Slope, Curved 4 x 2 No Studs
3066	Transparent	1	Brick 1 x 4 without Bottom Tubes
3023	Grey	9	Plate 1 x 2
87580	Grey	7	Plate, Modified 2 x 2 with Groove
4162	Grey	8	Tile 1 x 8
88072	Grey	1	Plate, Modified 1 x 2 with Arm Up

BLink ID	Color	Qty	Description
63864	Grey	4	Tile 1 x 3
3794	Grey	5	Plate, Modified 1 x 2 with 1 Stud (Jumper)
4592c02	Grey	1	Lever Small Base with Black Lever
60478	Grey	10	Plate, Modified 1 x 2 with Handle on End
4085c	Dark Grey	8	Plate, Modified 1 x 1 with Clip Vertical
4477	Grey	2	Plate 1 x 10
3068b	Grey	10	Tile 2 x 2 with Groove
6636	Grey	2	Tile 1 x 6
3021	Grey	2	Plate 2 x 3
3673	Grey	8	Technic, Pin without Friction
87407	Grey	8	Technic, Gear 20 Tooth Bevel with Pin Hole
3749	Grey	4	Technic, Axle Pin without Friction
4019	Grey	4	Technic, Gear 16 Tooth
60470	Grey	4	Plate, Modified 1 x 2 with Clips Horizontal
15712	Grey	5	Tile, Modified 1 x 1 with Clip
87087	Grey	6	Brick, Modified 1 x 1 with Stud on 1 Side
3070b	Transparent	2	Tile 1 x 1 with Groove
2412b	Dark Grey	3	Tile, Modified 1 x 2 Grille
3711	Black	82	Technic, Link Chain
48336	Grey	6	Plate, Modified 1 x 2 with Handle on Side
63965	Grey	1	Bar 6L with Stop Ring
4274	Grey	3	Technic, Pin 1/2
4073	Grey	2	Plate, Round 1 x 1 Straight Side
4735	Grey	3	Bar 1 x 3 with Clip and Stud Receptacle
30162	Dark Grey	1	Minifig, Utensil Binoculars Town
30374	Grey	5	Bar 4L (Lightsaber Blade / Wand)
30377	Grey	1	Arm Mechanical, Battle Droid
60474	Grey	1	Plate, Round 4 x 4 with Hole
2420	Grey	1	Plate 2 x 2 Corner
4032b	Dark Grey	1	Plate, Round 2 x 2 with Axle Hole
2540	Grey	1	Plate, Modified 1 x 2 with Handle on Side
3024	Grey	1	Plate 1 x 1

BLink ID	Color	Qty	Description
54200	Grey	2	Slope 30 1 x 1 x 2/3
4697b	Grey	1	Pneumatic T Piece New Style (T Bar)
2432	Grey	1	Tile, Modified 1 x 2 with Handle
63868	Grey	1	Plate, Modified 1 x 2 with Clip
3794	Brown	1	Plate, Modified 1 x 2 with 1 Stud (Jumper)
30663	Black	1	Vehicle, Steering Wheel Small
4733	Grey	1	Brick, Modified 1 x 1 with Studs on 4 Sides
4740	Grey	2	Dish 2 x 2 Inverted (Radar)

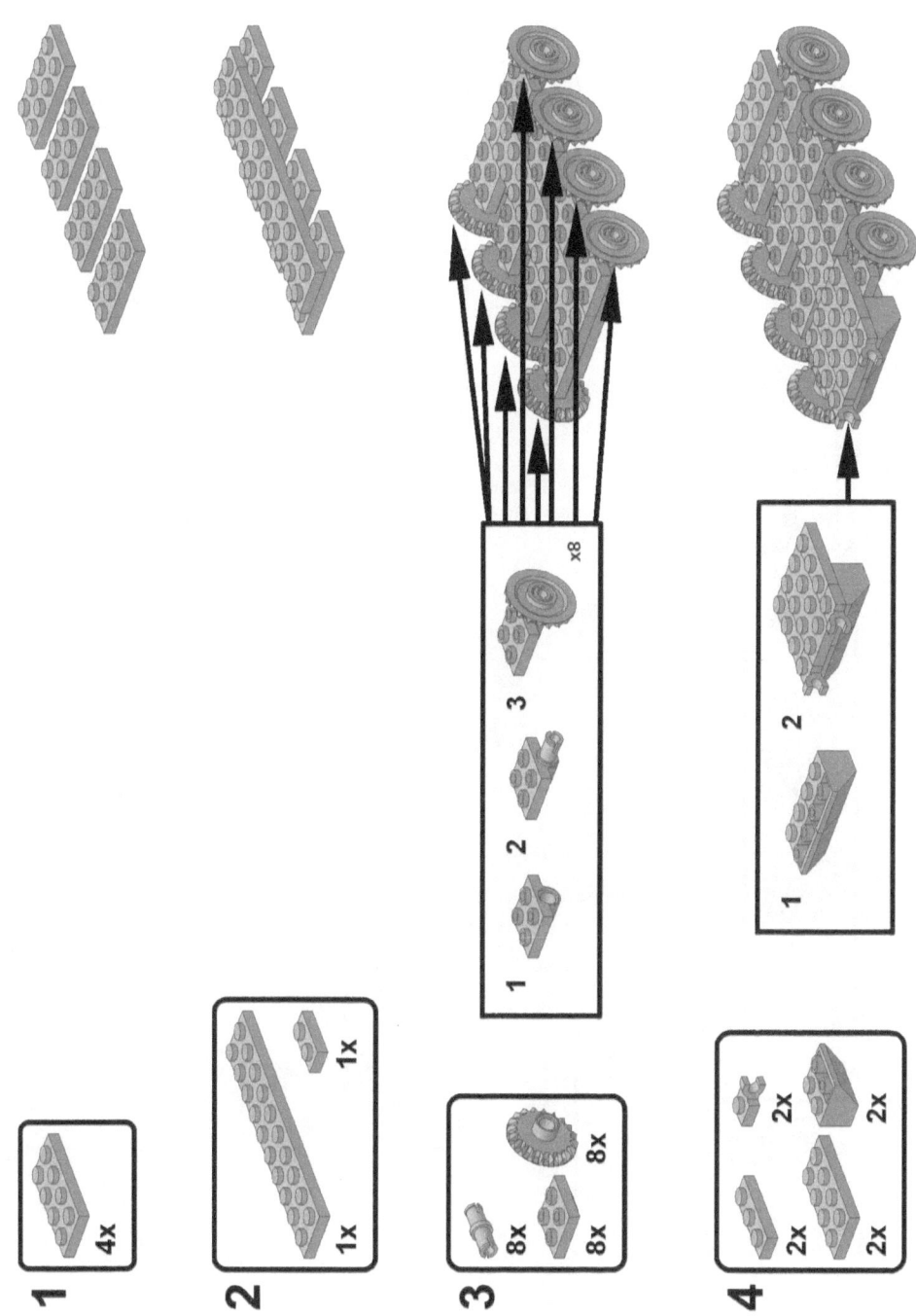

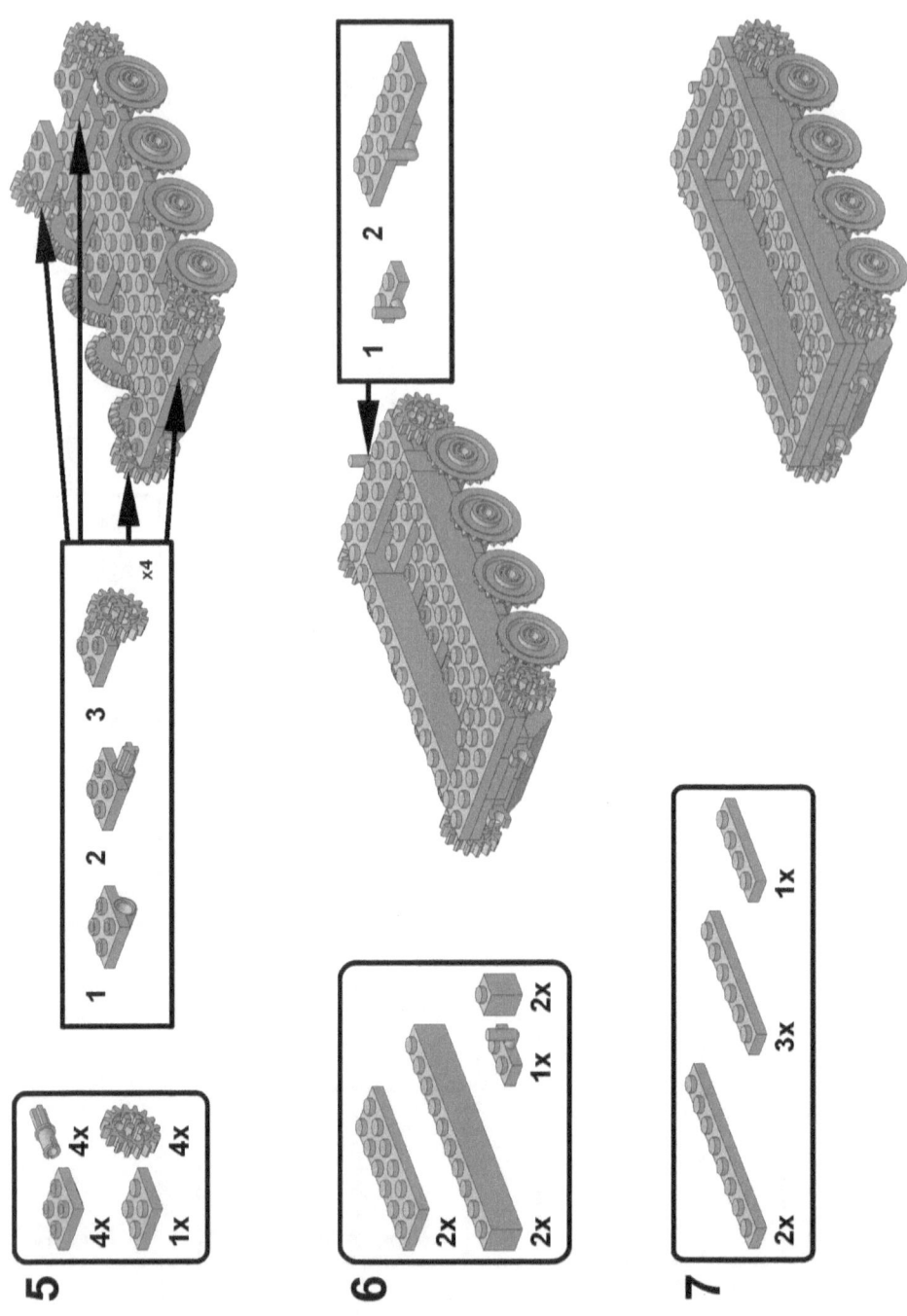

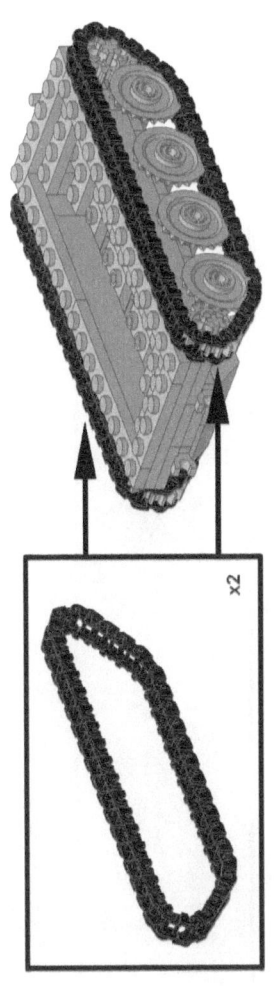
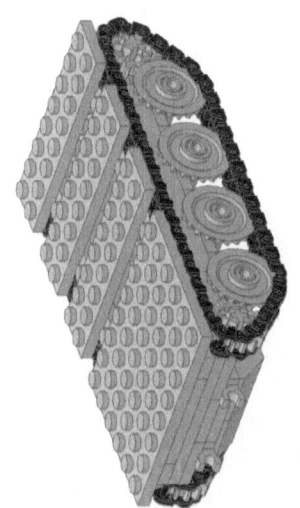
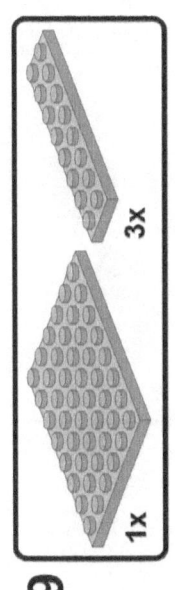

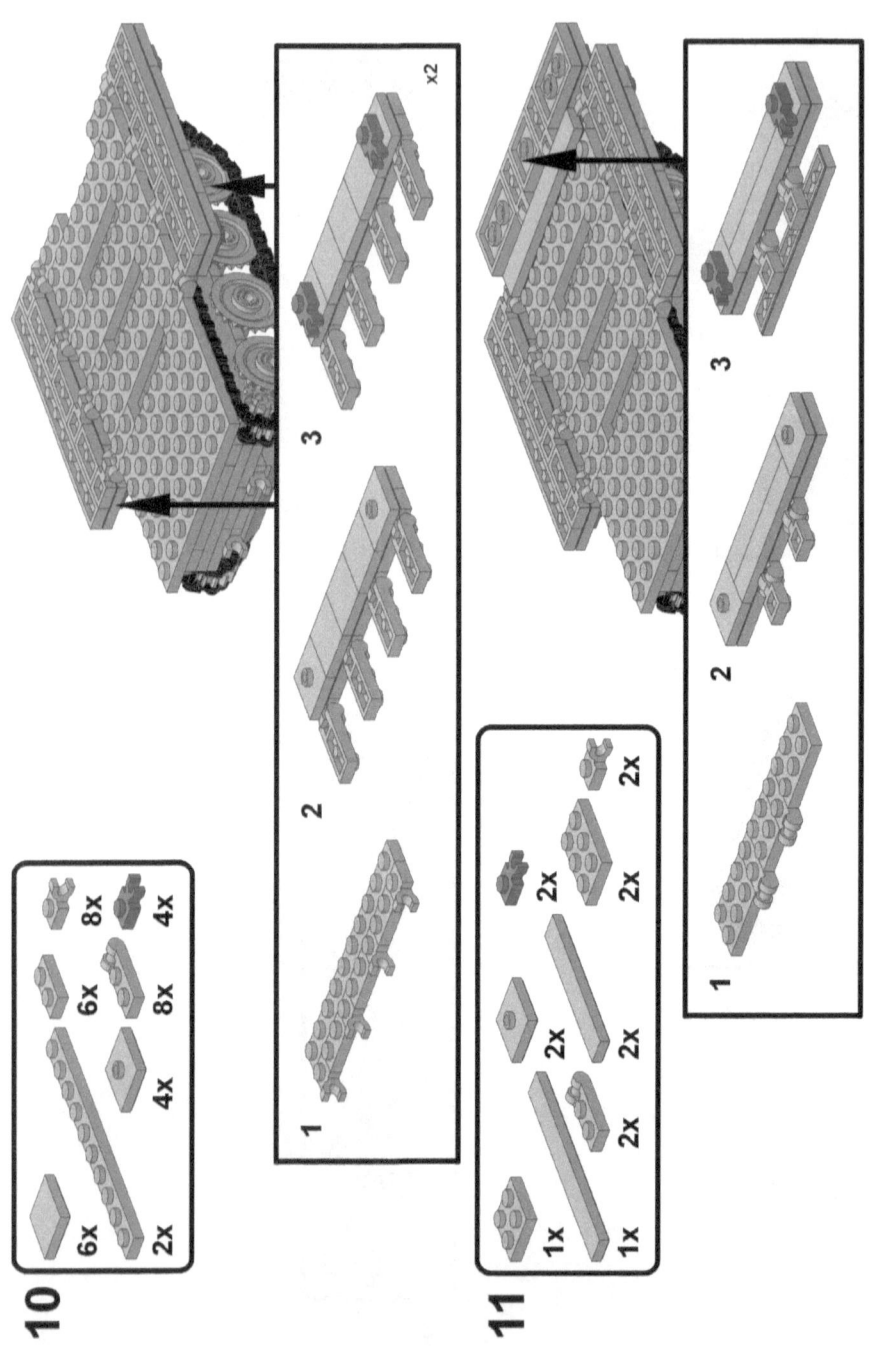

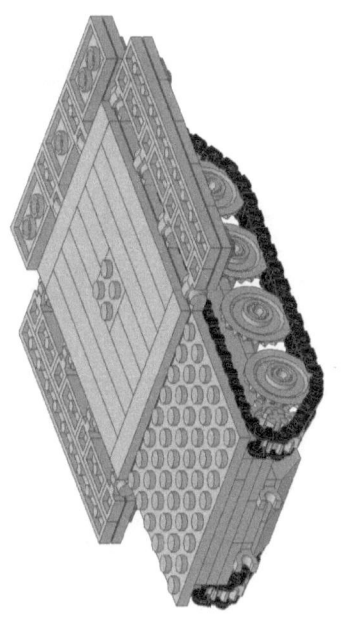

12
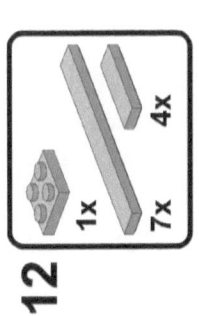
1x, 7x, 4x

13
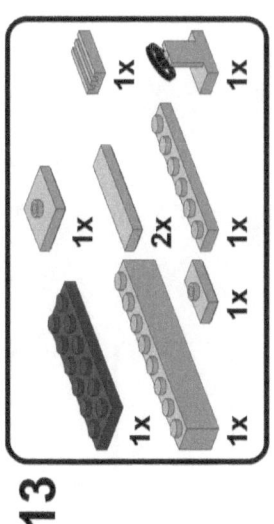
1x, 1x, 1x, 1x, 2x, 1x, 1x, 1x, 1x

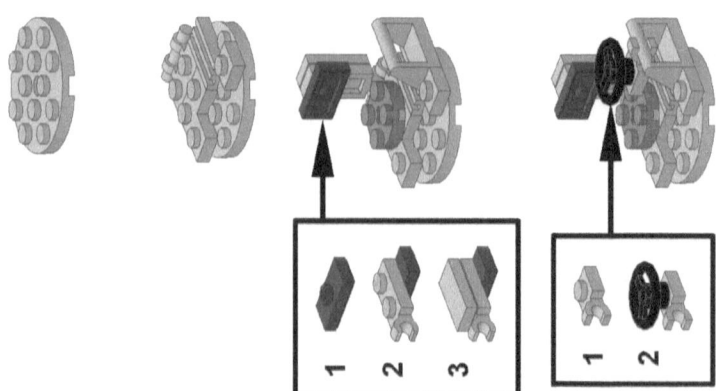
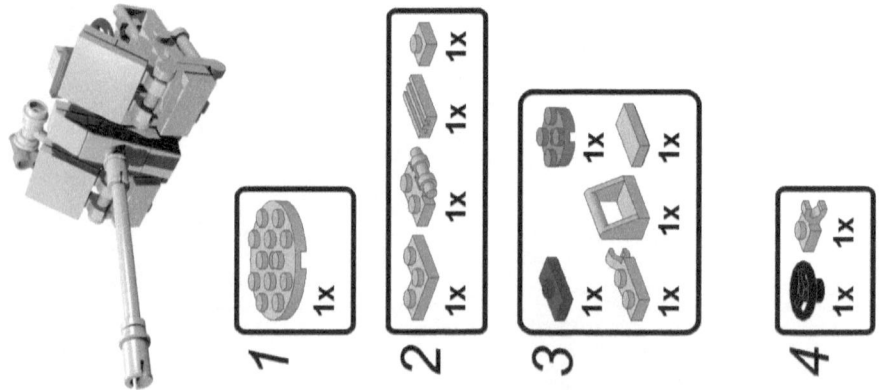

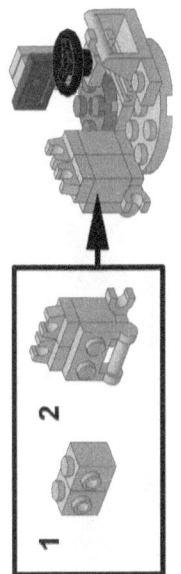
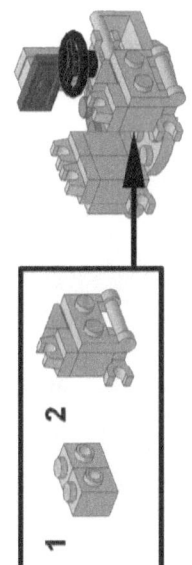
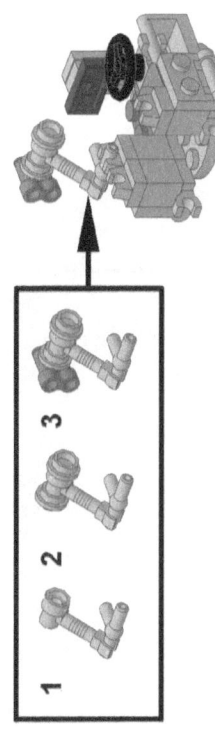
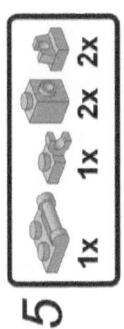
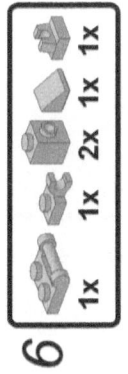

www.ww2custombrickmodels.de

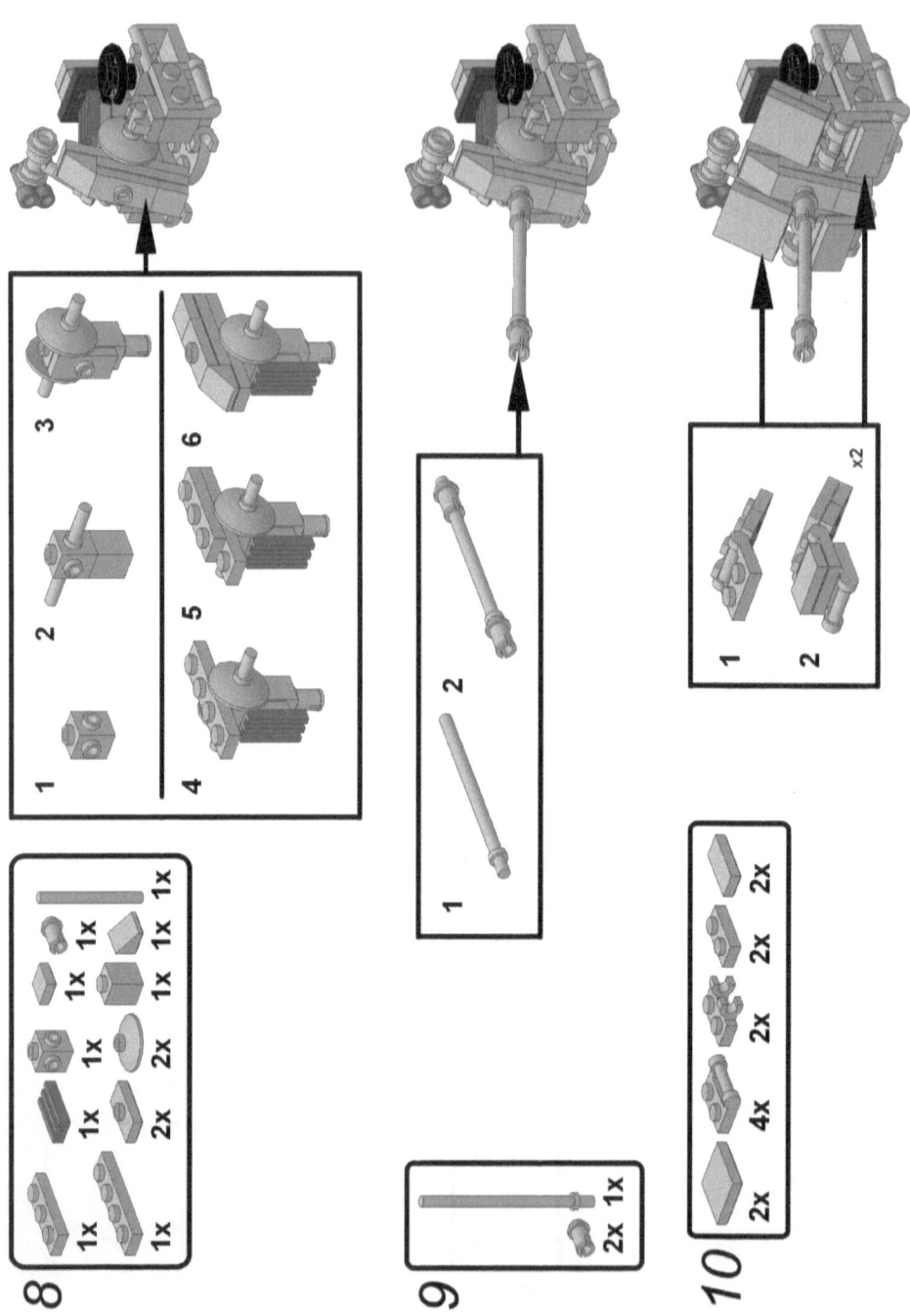

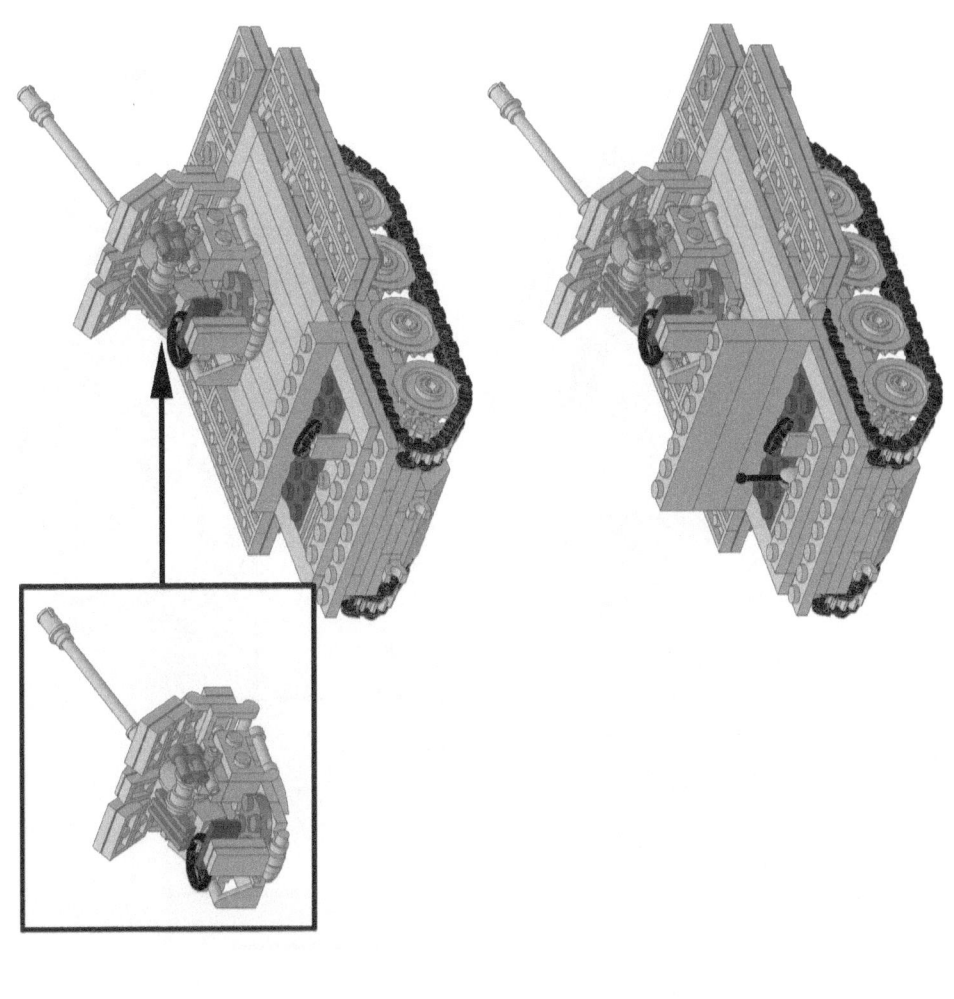

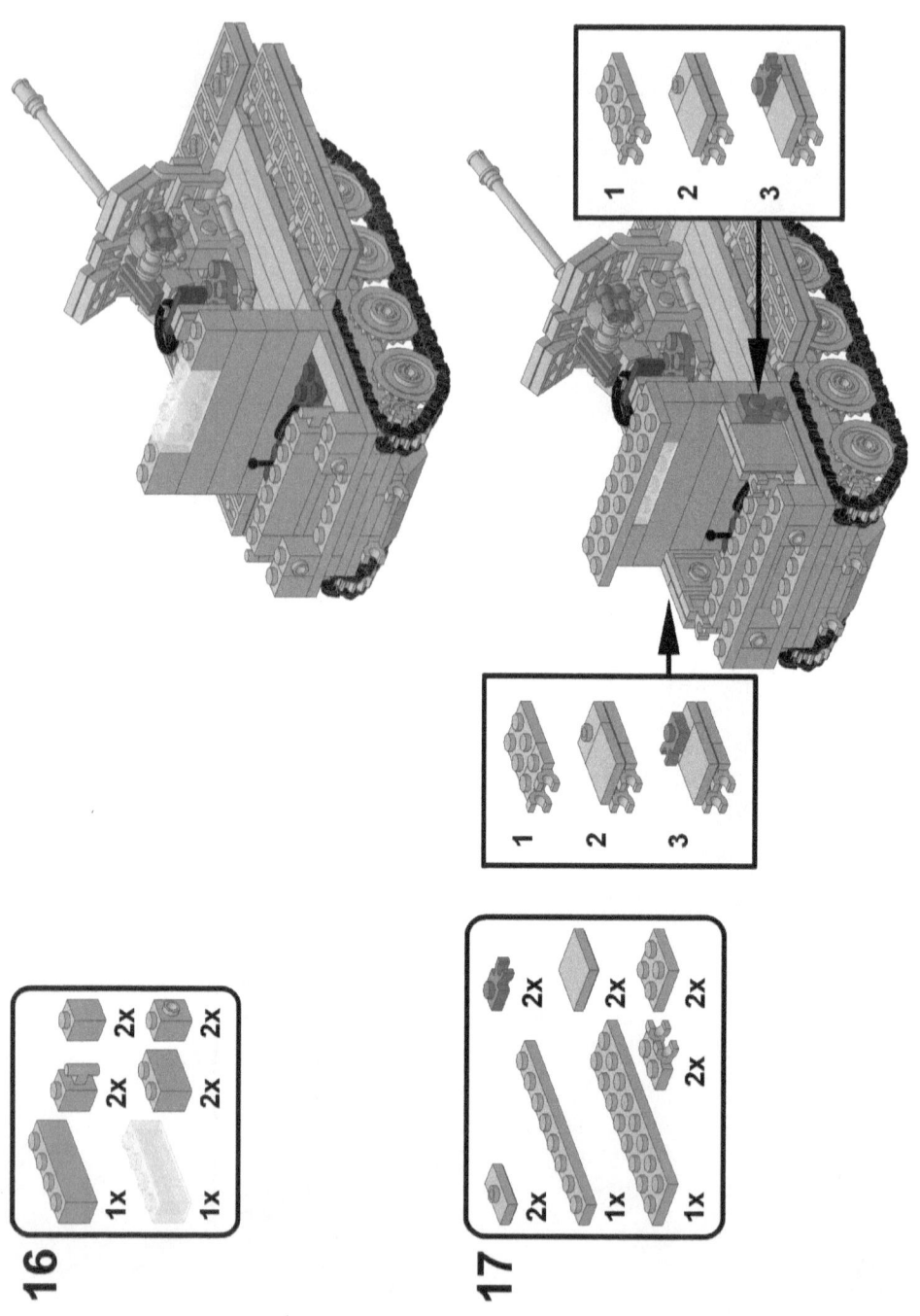

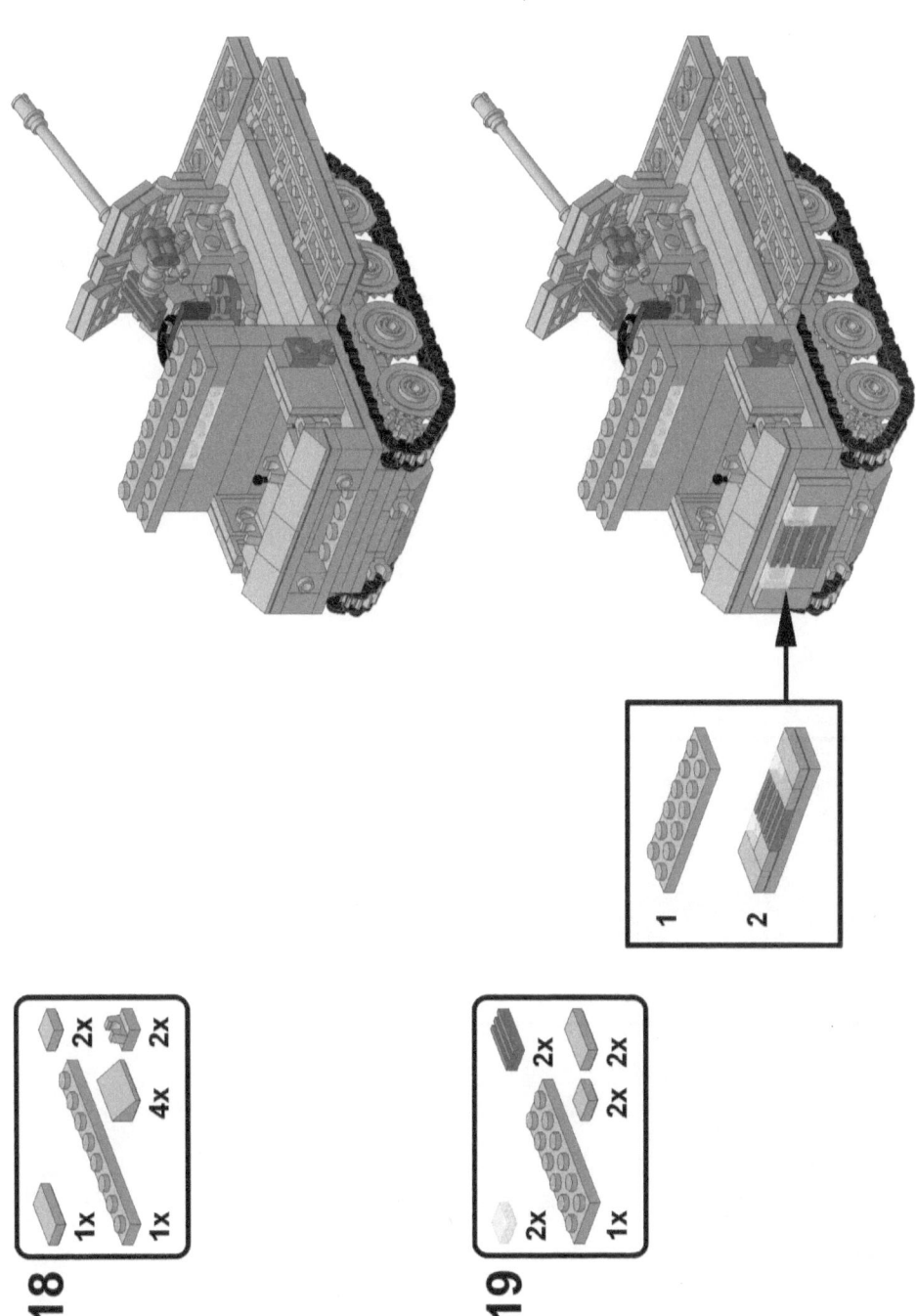

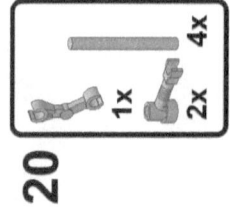

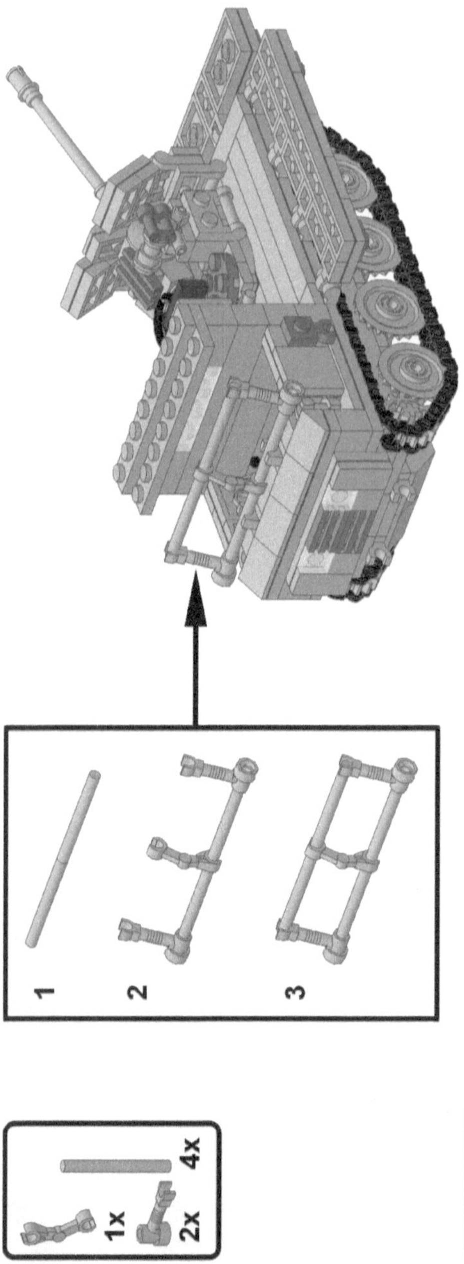

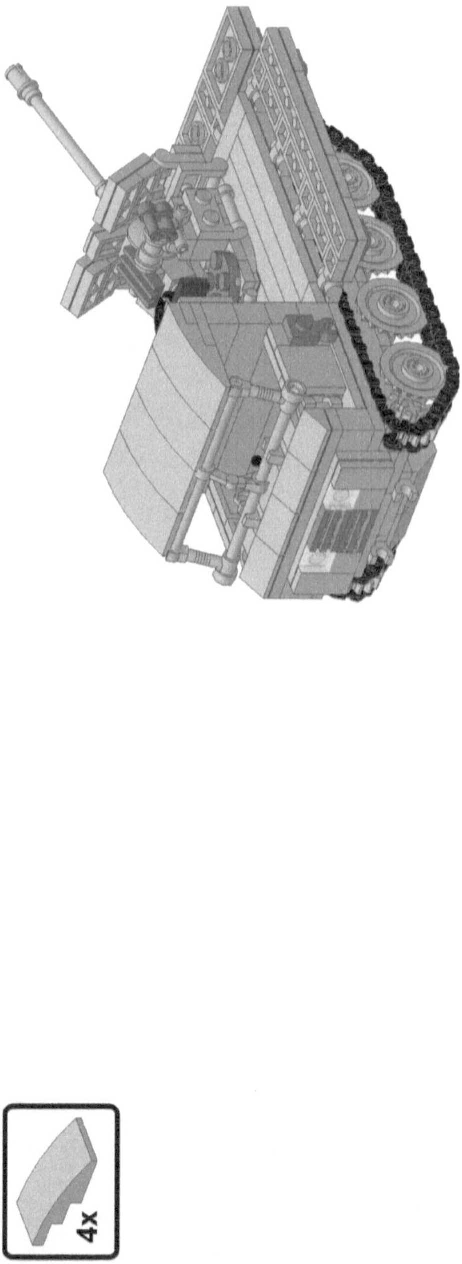

𝔖𝔇𝔎𝔉ℨ 250

Zum Bau des Modells benötigen Sie ca. 292 LEGO® Bausteine.
Länge: ca. 13,7 cm Breite: ca. 6,4 cm Höhe: ca. 6,7 cm

𝔖𝔇𝔎𝔉ℨ 250

Requires approx. 292 LEGO® bricks.
length: ca. 13.7 cm width: ca. 8.4 cm height: ca. 6.7 cm

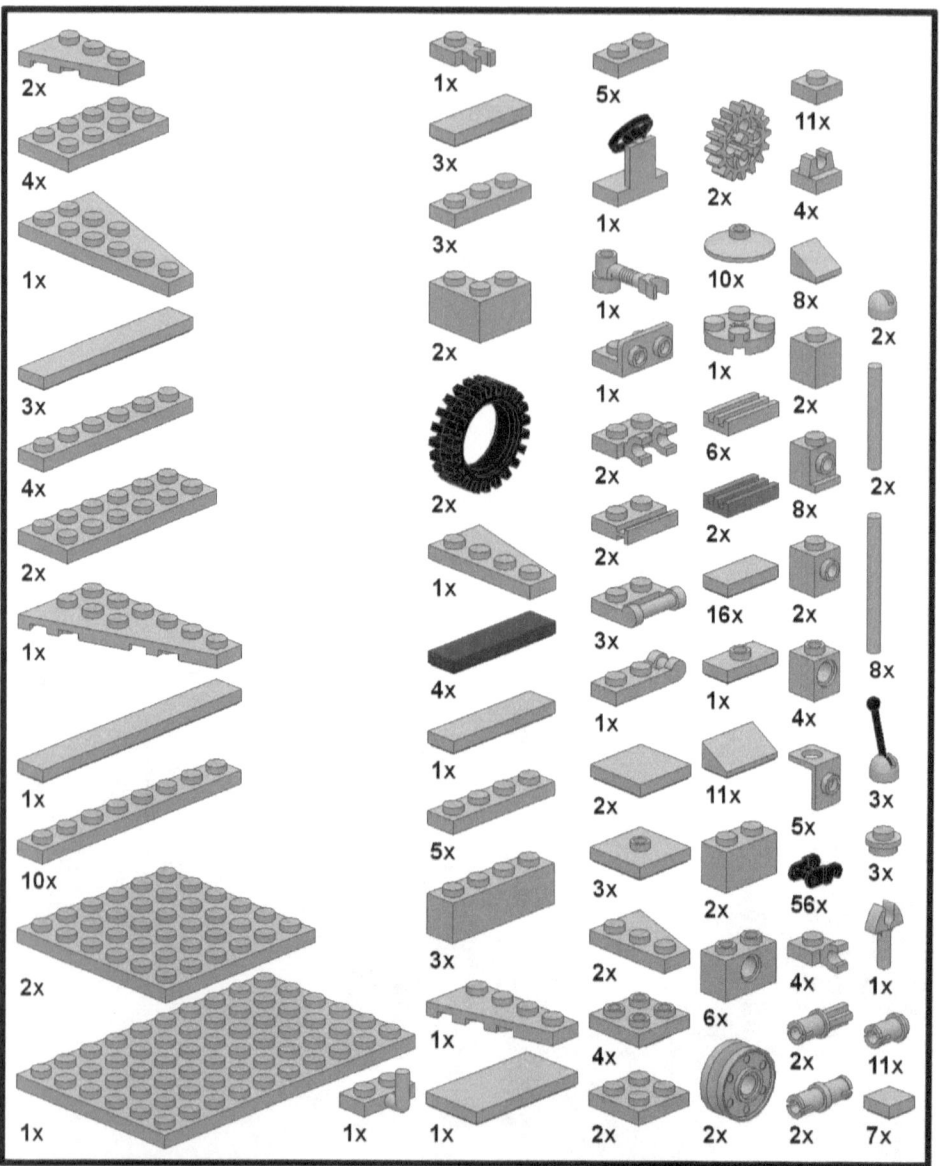

BLink ID	Color	Qty	Description
63864	Grey	3	Tile 1 x 3
41770	Grey	1	Wedge, Plate 4 x 2 Left
41769	Grey	1	Wedge, Plate 4 x 2 Right

BLink ID	Color	Qty	Description
3829c01	Black	1	Vehicle, Steering Stand 1 x 2
3033	Grey	1	Plate 6 x 10
3958	Grey	2	Plate 6 x 6
3020	Grey	4	Plate 2 x 4
3623	Grey	3	Plate 1 x 3
47905	Grey	2	Brick, Modified 1 x 1 with Studs on 2 Sides
54200	Grey	8	Slope 30 1 x 1 x 2/3
3068b	Grey	2	Tile 2 x 2 with Groove
4162	Grey	1	Tile 1 x 8
3010	Grey	3	Brick 1 x 4
3460	Grey	10	Plate 1 x 8
3069b	Grey	16	Tile 1 x 2 with Groove
85984	Grey	11	Slope 30 1 x 2 x 2/3
2357	Grey	2	Brick 2 x 2 Corner
3024	Grey	11	Plate 1 x 1
60470	Grey	2	Plate, Modified 1 x 2 with Clips Horizontal
3070b	Grey	7	Tile 1 x 1 with Groove
3023	Grey	5	Plate 1 x 2
4073	Grey	3	Plate, Round 1 x 1 Straight Side
2555	Grey	4	Tile, Modified 1 x 1 with Clip
42446	Grey	5	Minifig, Neck Bracket with Back Stud
2444	Grey	4	Plate, Modified 2 x 2 with Pin Hole
6019	Grey	4	Plate, Modified 1 x 1 with Clip Horizontal
15573	Grey	1	Plate, Modified 1 x 2 with 1 Stud
4592c02	Grey	5	Lever Small Base with Black Lever
3710	Grey	5	Plate 1 x 4
99780	Grey	1	Bracket 1 x 2 - 1 x 2 Inverted
2412b	Grey	6	Tile, Modified 1 x 2 Grille
32028	Grey	2	Plate, Modified 1 x 2 with Door Rail
2431	Brown	4	Tile 1 x 4
4032b	Grey	1	Plate, Round 2 x 2 with Axle Hole
3700	Grey	6	Technic, Brick 1 x 2 with Hole

BLink ID	Color	Qty	Description
3795	Grey	2	Plate 2 x 6
4070	Grey	8	Brick, Modified 1 x 1 with Headlight
3004	Grey	2	Brick 1 x 2
3005	Grey	2	Brick 1 x 1
6541	Grey	4	Technic, Brick 1 x 1 with Hole
3666	Grey	4	Plate 1 x 6
88072	Grey	1	Plate, Modified 1 x 2 with Arm Up
48336	Grey	3	Plate, Modified 1 x 2 with Handle on Side
4274	Grey	11	Technic, Pin 1/2
30374	Grey	8	Bar 4L (Lightsaber Blade / Wand)
4740	Grey	10	Dish 2 x 2 Inverted (Radar)
3749	Grey	2	Technic, Axle Pin without Friction
4019	Grey	2	Technic, Gear 16 Tooth
87580	Grey	3	Plate, Modified 2 x 2 with Groove and 1 Stud
87994	Grey	2	Bar 3L
3673	Grey	2	Technic, Pin without Friction Ridges Lengthwise
56902	Grey	2	Wheel 18mm D. x 8mm
61254	Black	2	Tire Offset Tread - Band Around Center of Tread
43722	Grey	2	Wedge, Plate 3 x 2 Right
43723	Grey	2	Wedge, Plate 3 x 2 Left
3711	Black	56	Technic, Link Chain
87079	Grey	1	Tile 2 x 4
54384	Grey	1	Wedge, Plate 6 x 3 Left
6636	Grey	3	Tile 1 x 6
54383	Grey	1	Wedge, Plate 6 x 3 Right
2412b	Dark Grey	2	Tile, Modified 1 x 2 Grille
3022	Grey	2	Plate 2 x 2
2431	Grey	1	Tile 1 x 4
60478	Grey	1	Plate, Modified 1 x 2 with Handle on End
4085c	Grey	1	Plate, Modified 1 x 1 with Clip Vertical
4735	Grey	1	Bar 1 x 3 with Clip and Stud Receptacle
48729b	Grey	1	Bar 1L with Clip Mechanical Claw

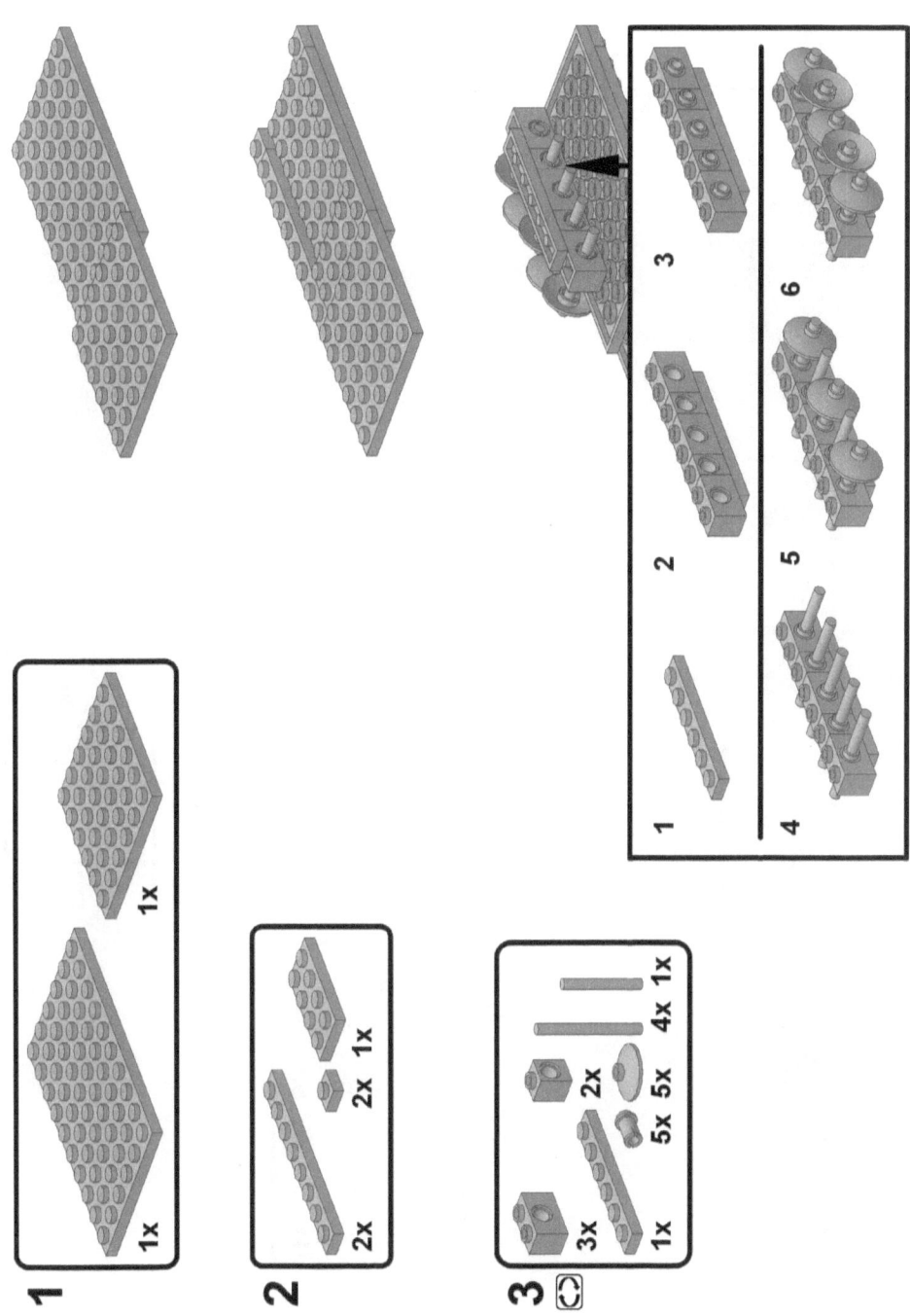

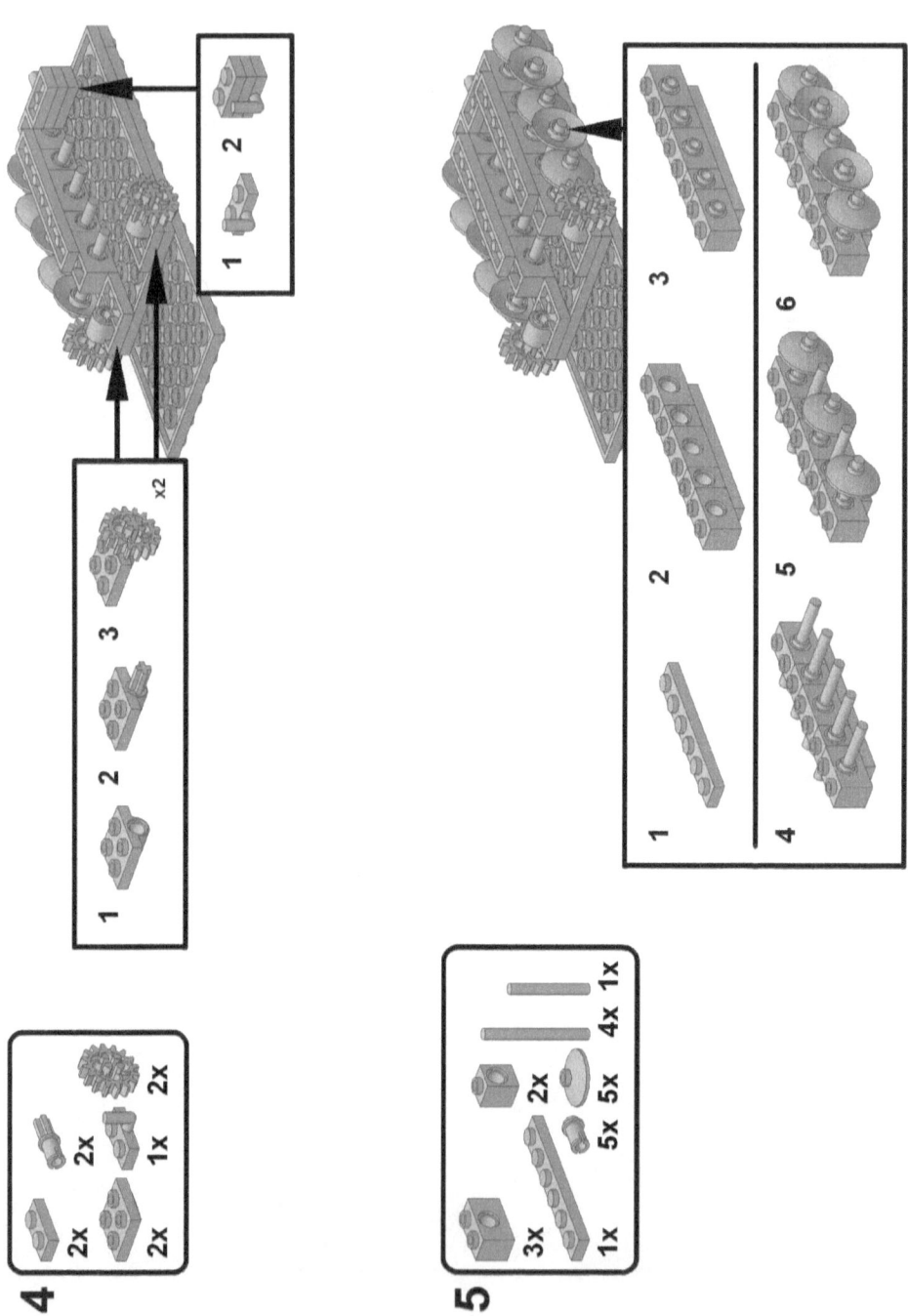

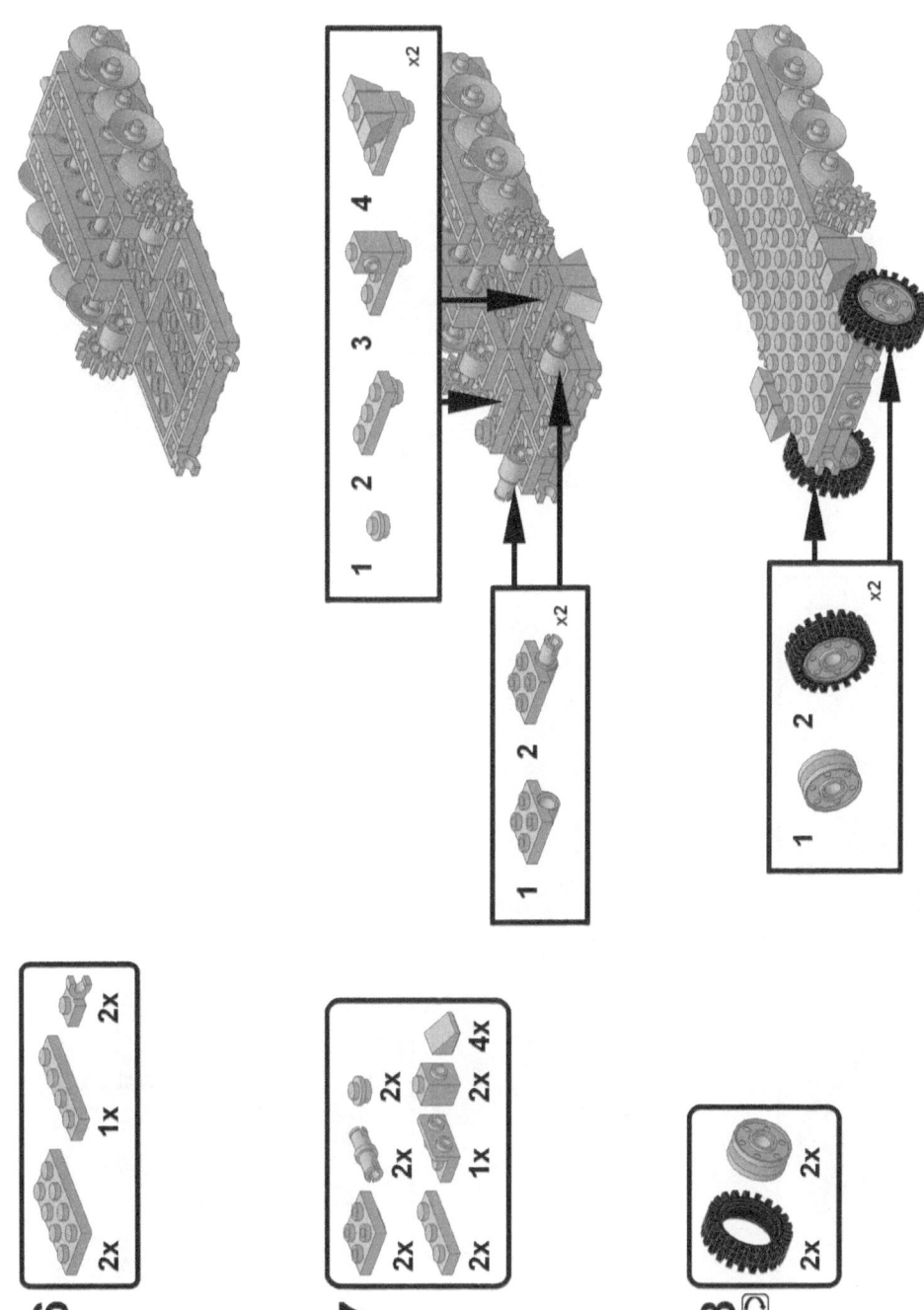

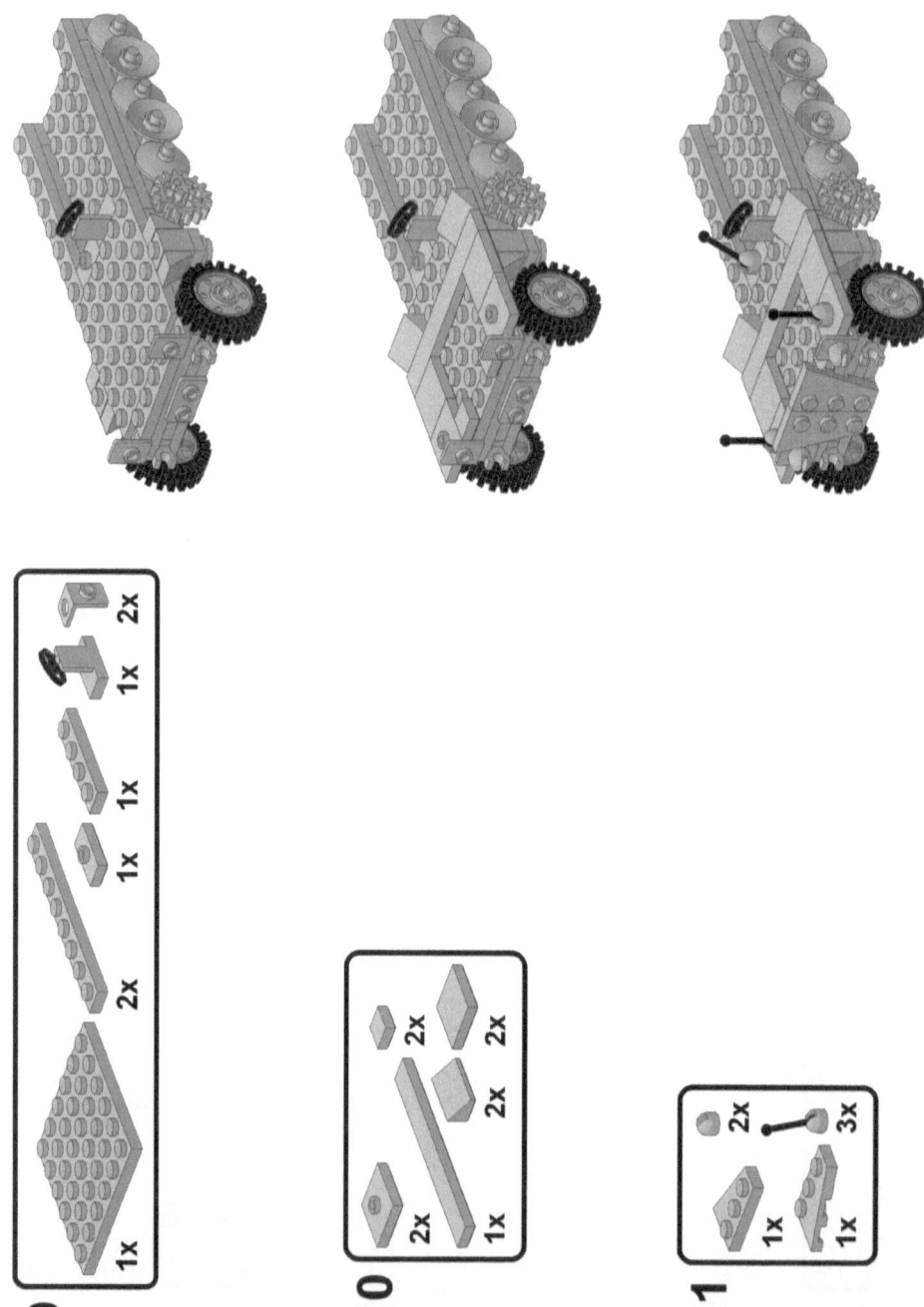

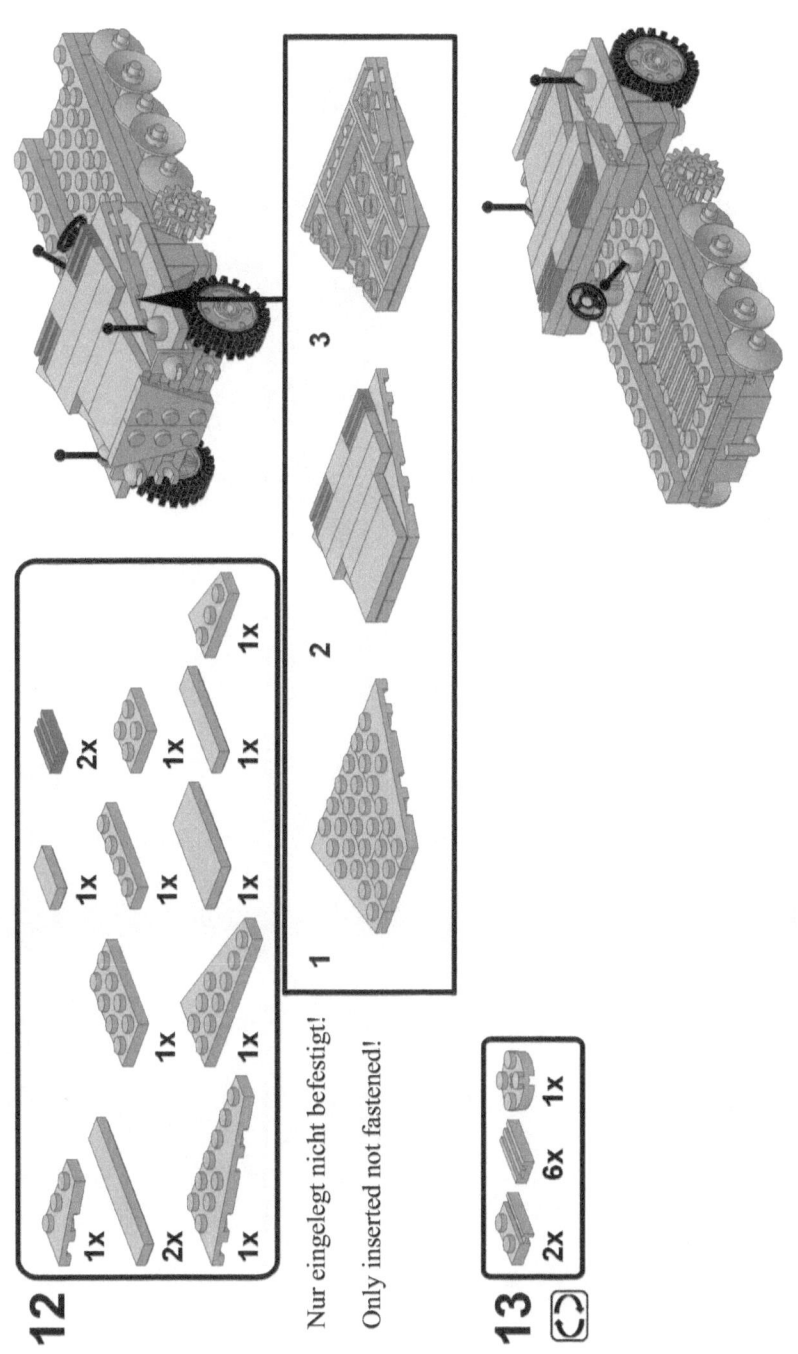

Nur eingelegt nicht befestigt!
Only inserted not fastened!

www.ww2custombrickmodels.de

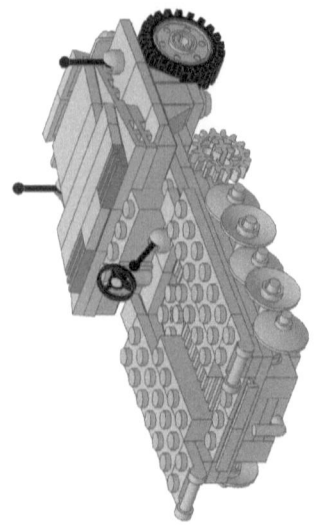
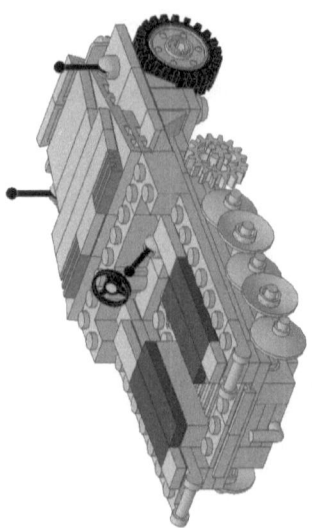
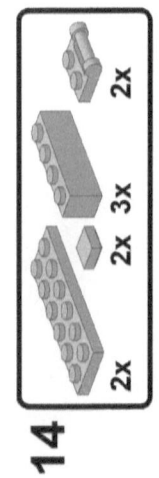
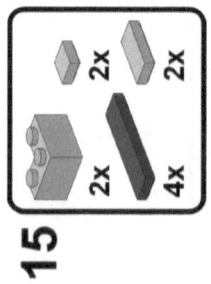

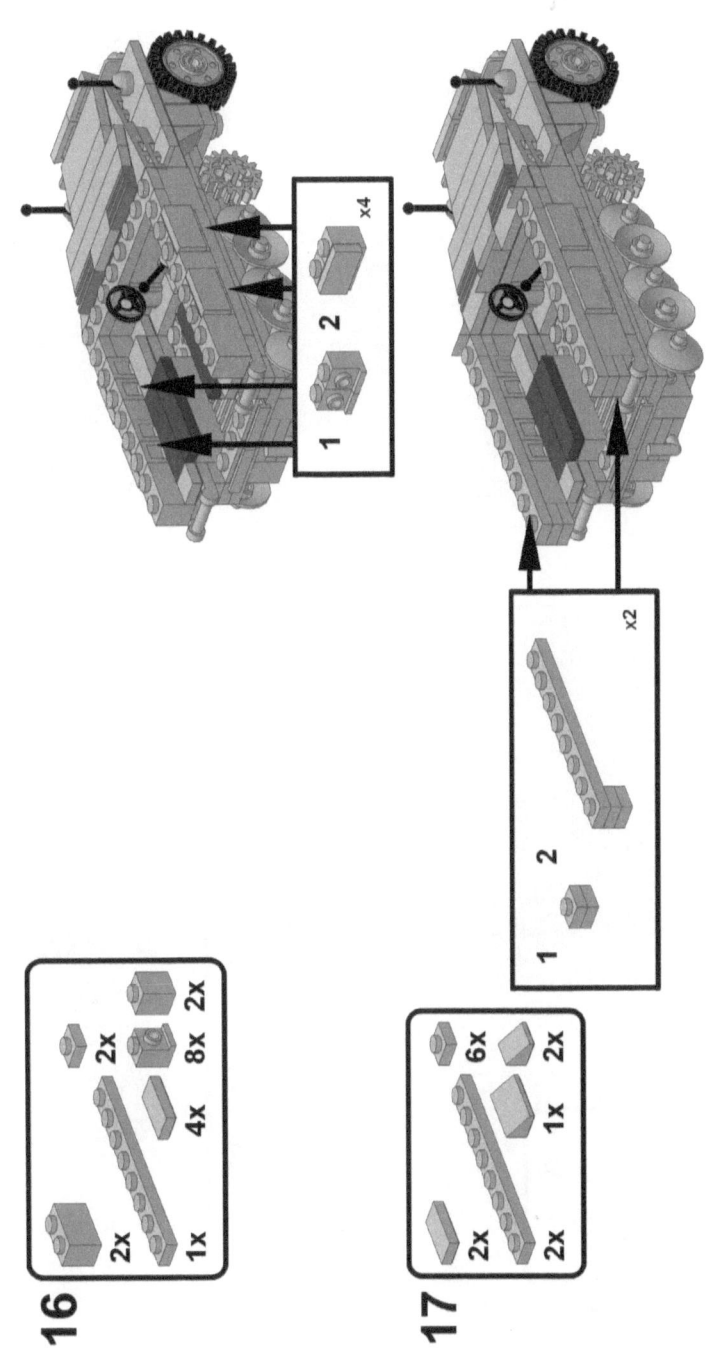

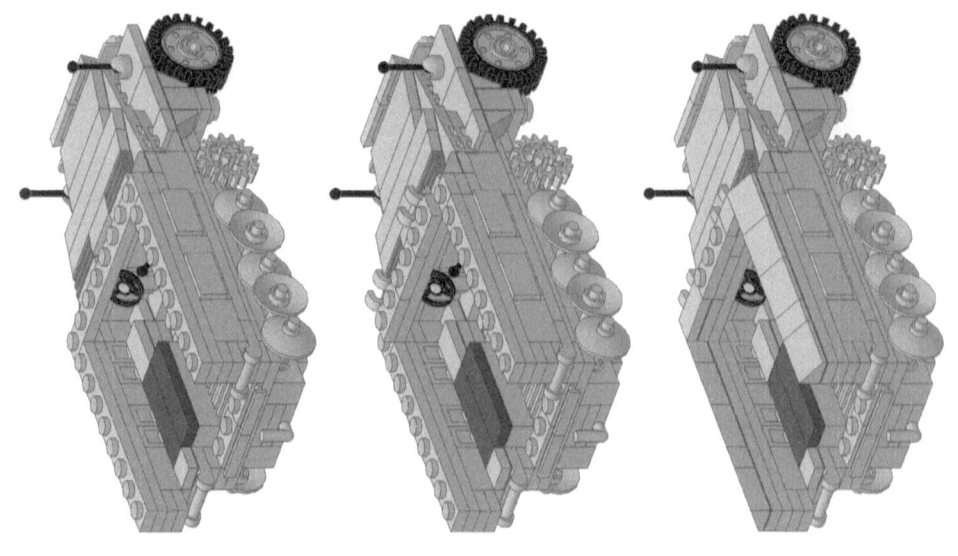

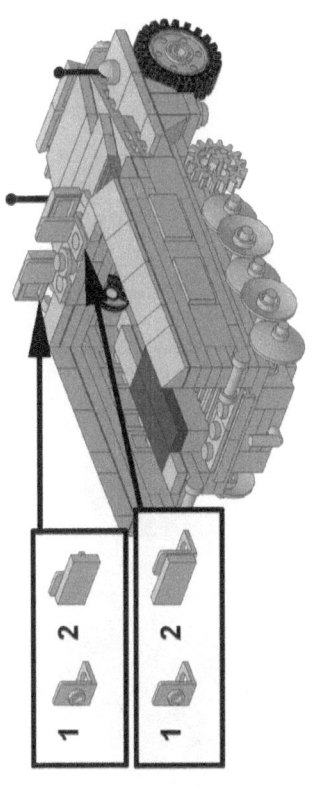
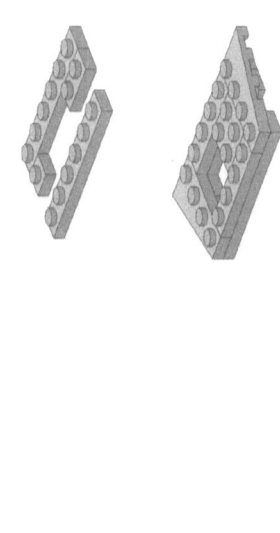
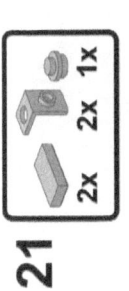
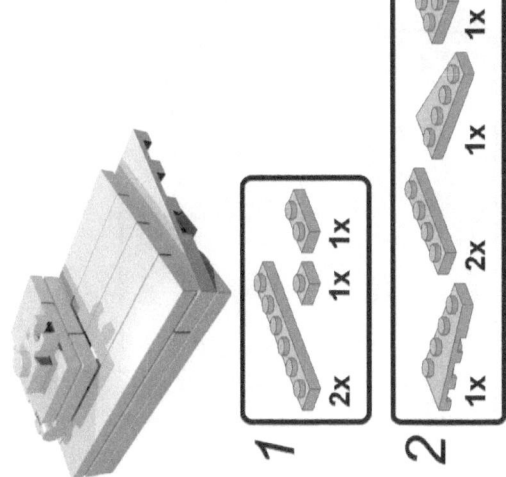

www.ww2custombrickmodels.de

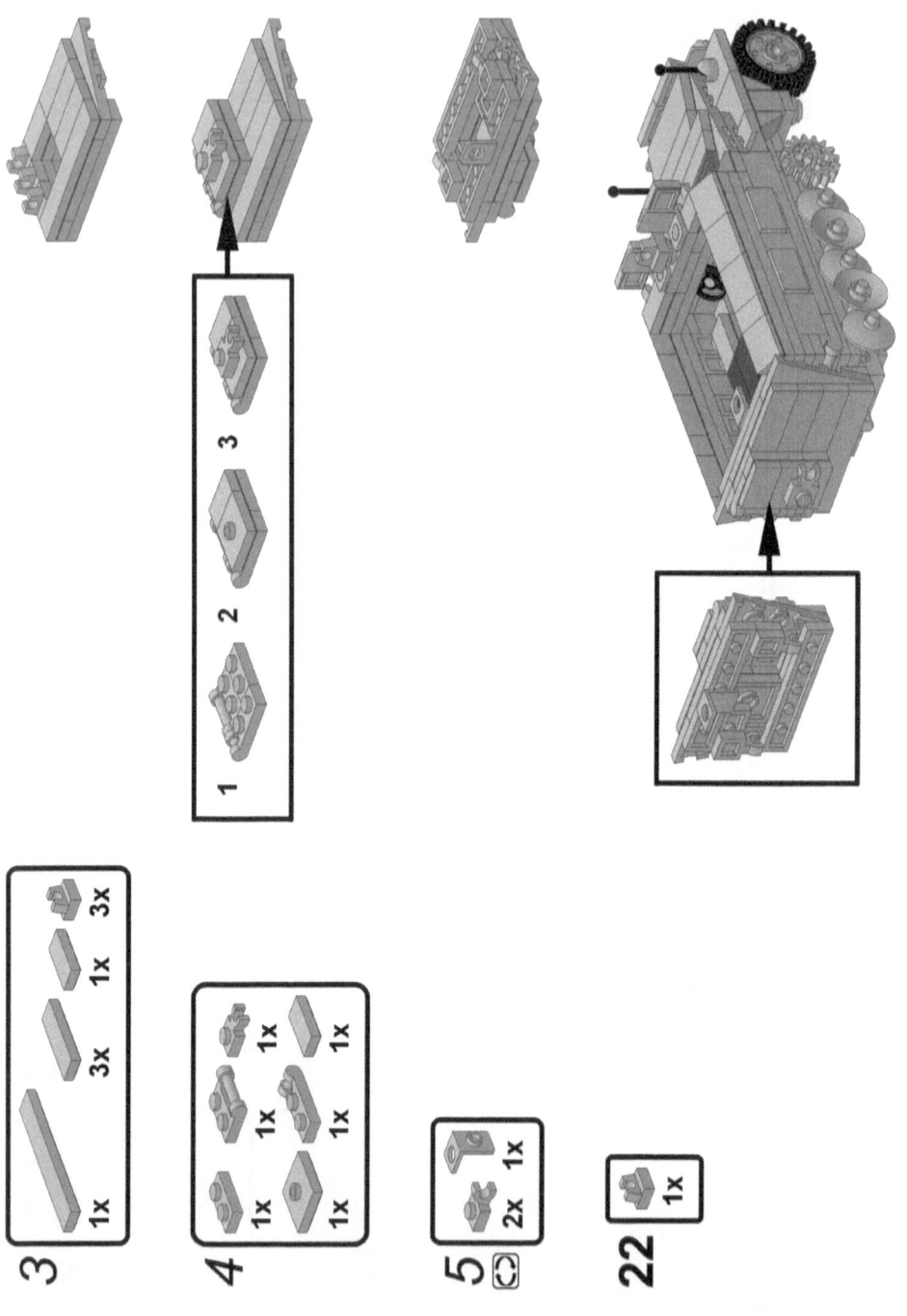

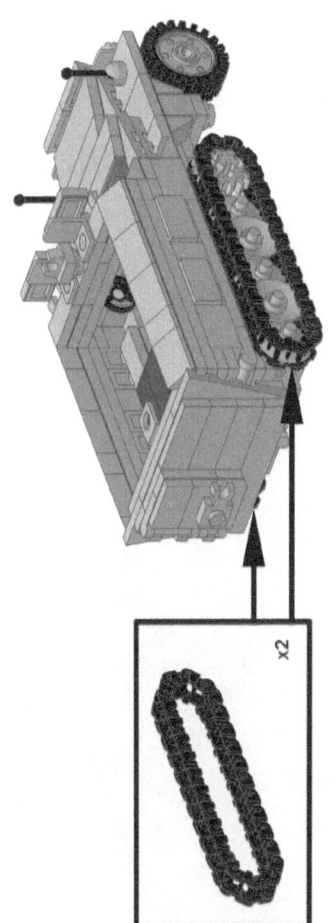
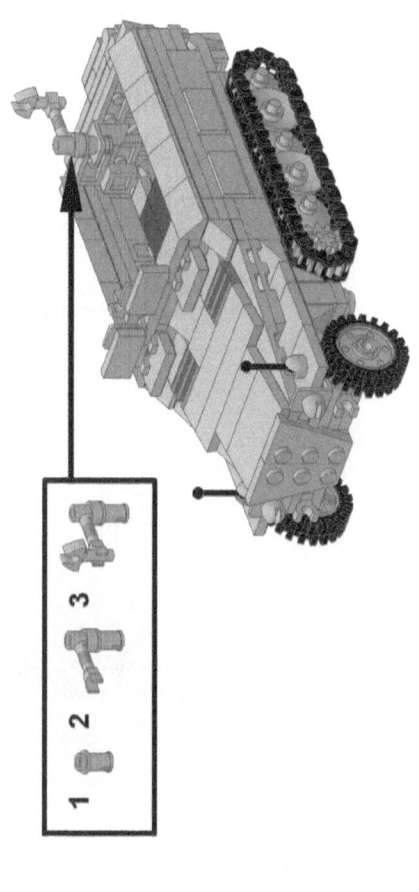

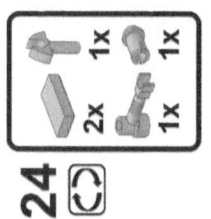

Stug 3

Zum Bau des Modells benötigen Sie ca. 440 LEGO® Bausteine.
Länge: ca. 17,2 cm Breite: ca. 9,6 cm Höhe: ca. 7,2 cm

Stug 3

Requires approx. 440 LEGO® bricks.
length: ca. 17.2 cm width: ca. 9.6 cm height: ca. 7.2 cm

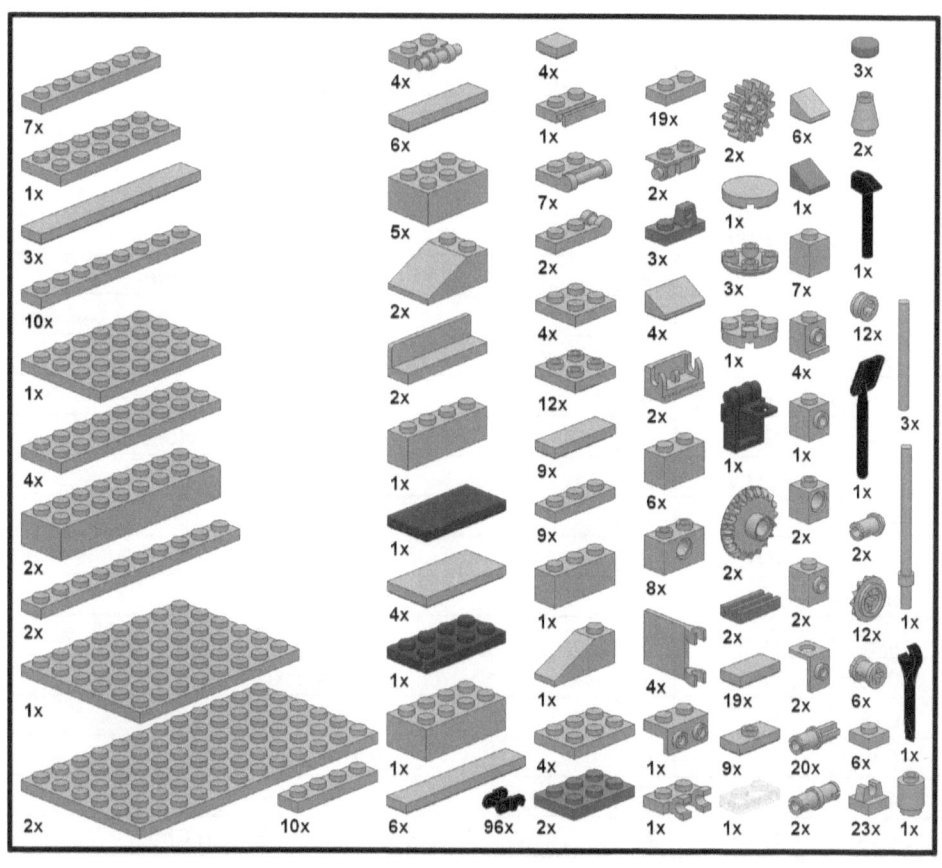

BLink ID	Color	Qty	Description
87407	Grey	2	Technic, Gear 20 Tooth Bevel with Pin Hole
3673	Grey	2	Technic, Pin without Friction
3032	Grey	1	Plate 4 x 6
2444	Grey	12	Plate, Modified 2 x 2 with Pin Hole
3460	Grey	10	Plate 1 x 8
6541	Grey	2	Technic, Brick 1 x 1 with Hole
3023	Grey	19	Plate 1 x 2
3666	Grey	7	Plate 1 x 6
3034	Grey	4	Plate 2 x 8
3700	Grey	8	Technic, Brick 1 x 2 with Hole
3710	Grey	10	Plate 1 x 4
4477	Grey	2	Plate 1 x 10
3623	Grey	9	Plate 1 x 3
3036	Grey	1	Plate 6 x 8
3004	Grey	6	Brick 1 x 2
3005	Grey	7	Brick 1 x 1
3028	Grey	2	Plate 6 x 12
6636	Grey	6	Tile 1 x 6
2431	Grey	6	Tile 1 x 4
2555	Grey	23	Tile, Modified 1 x 1 with Clip
63864	Grey	9	Tile 1 x 3
4162	Grey	3	Tile 1 x 8
3024	Grey	6	Plate 1 x 1
54200	Grey	6	Slope 30 1 x 1 x 2/3
3069b	Grey	19	Tile 1 x 2 with Groove
3001	Grey	1	Brick 2 x 4
3070b	Grey	4	Tile 1 x 1 with Groove
3007	Grey	2	Brick 2 x 8
3002	Grey	5	Brick 2 x 3
30413	Grey	2	Panel 1 x 4 x 1
3794	Grey	9	Plate, Modified 1 x 2 with 1 Stud (Jumper)
2412b	Dark Grey	2	Tile, Modified 1 x 2 Grille

BLink ID	Color	Qty	Description
3023	Transparent	1	Plate 1 x 2
32028	Grey	1	Plate, Modified 1 x 2 with Door Rail
3022	Grey	4	Plate 2 x 2
4070	Grey	4	Brick, Modified 1 x 1 with Headlight
3937	Grey	2	Hinge Brick 1 x 2 Base
3795	Grey	1	Plate 2 x 6
3021	Grey	4	Plate 2 x 3
48336	Grey	7	Plate, Modified 1 x 2 with Handle on Side
2654	Grey	3	Plate, Round 2 x 2 with Rounded Bottom
3298	Grey	2	Slope 33 3 x 2
3622	Grey	1	Brick 1 x 3
4286	Grey	1	Slope 33 3 x 1
3010	Grey	1	Brick 1 x 4
2540	Grey	4	Plate, Modified 1 x 2 with Handle on Side
3021	Dark Grey	2	Plate 2 x 3
30383	Dark Grey	3	Hinge Plate 1 x 2 Locking with 1 Finger On Top
4522	Black	1	Minifig, Utensil Tool Mallet / Hammer
3749	Grey	20	Technic, Axle Pin without Friction
94925	Grey	2	Technic, Gear 16 Tooth (New Style Reinforced)
3713	Grey	6	Technic Bush
4265c	Grey	12	Technic Bush 1/2 Smooth
6589	Grey	12	Technic, Gear 12 Tooth Bevel
3837	Black	1	Minifig, Utensil Shovel (Round Stem End)
98138	Dark Grey	3	Tile, Round 1 x 1
3938	Grey	2	Hinge Brick 1 x 2 Top Plate Thin
2335	Grey	4	Flag 2 x 2 Square
30374	Grey	3	Bar 4L (Lightsaber Blade / Wand)
4006	Black	1	Minifig, Utensil Tool Spanner / Screwdriver
87087	Grey	2	Brick, Modified 1 x 1 with Stud on 1 Side
3711	Black	96	Technic, Link Chain
87079	Grey	4	Tile 2 x 4
60470	Grey	1	Plate, Modified 1 x 2 with Clips Horizontal

BLink ID	Color	Qty	Description
99781	Grey	1	Bracket 1 x 2 - 1 x 2
85984	Grey	4	Slope 30 1 x 2 x 2/3
3020	Brown	1	Plate 2 x 4
87079	Brown	1	Tile 2 x 4
42446	Grey	2	Minifig, Neck Bracket with Back Stud
54200	Dark Grey	1	Slope 30 1 x 1 x 2/3
2524	Brown	1	Minifig, Backpack Non-Opening
60478	Grey	2	Plate, Modified 1 x 2 with Handle on End
4032b	Grey	1	Plate, Round 2 x 2 with Axle Hole
4150	Grey	1	Tile, Round 2 x 2
47905	Grey	1	Brick, Modified 1 x 1 with Studs on 2 Sides
3062b	Grey	1	Brick, Round 1 x 1 Open Stud
4589	Grey	2	Cone 1 x 1 without Top Groove
63965	Grey	1	Bar 6L with Stop Ring
4274	Grey	2	Technic, Pin 1/2

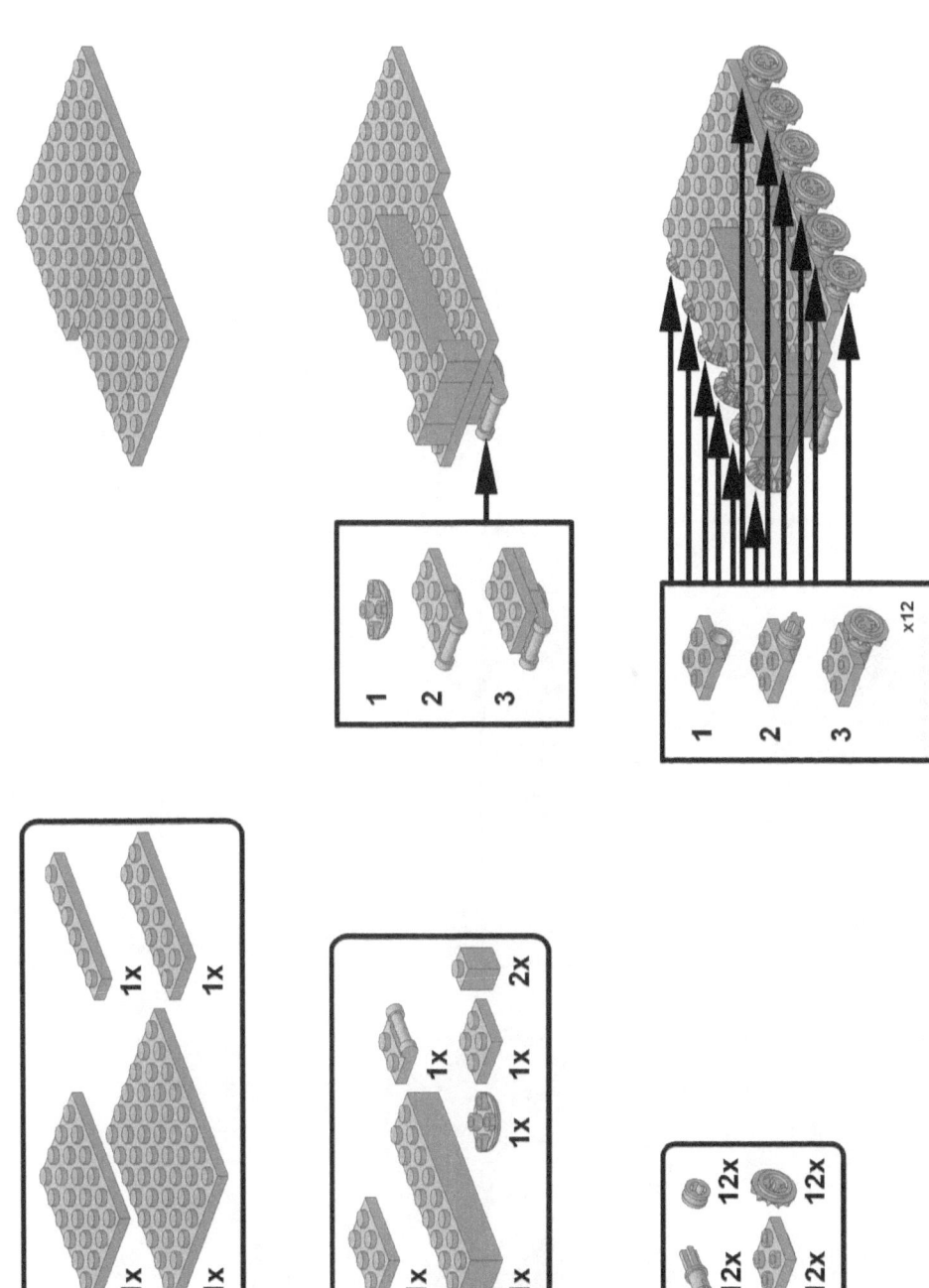

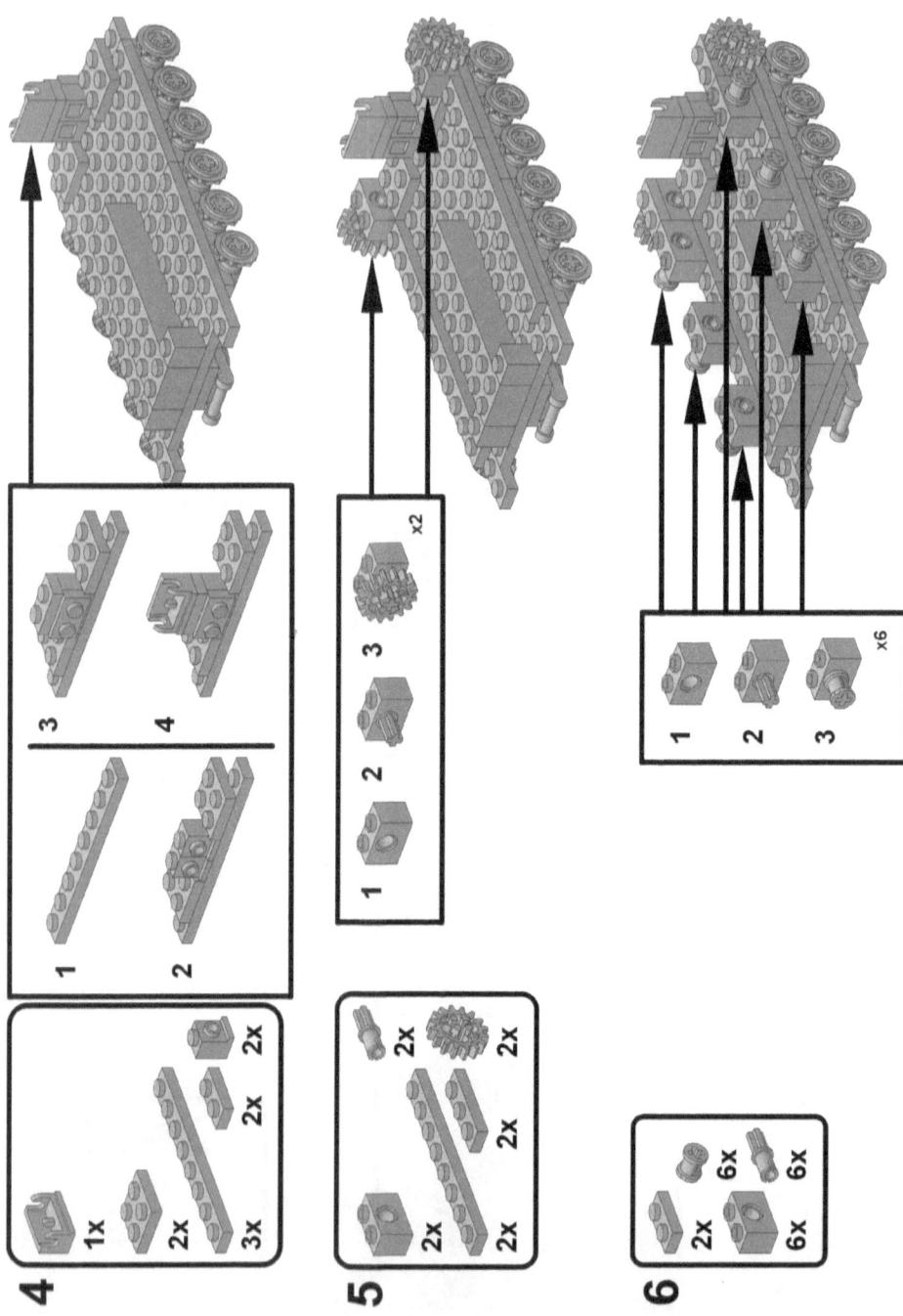

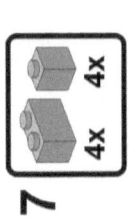
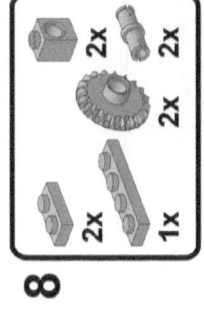
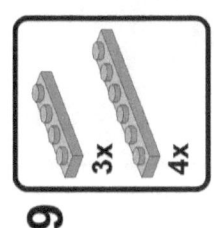

www.ww2custombrickmodels.de

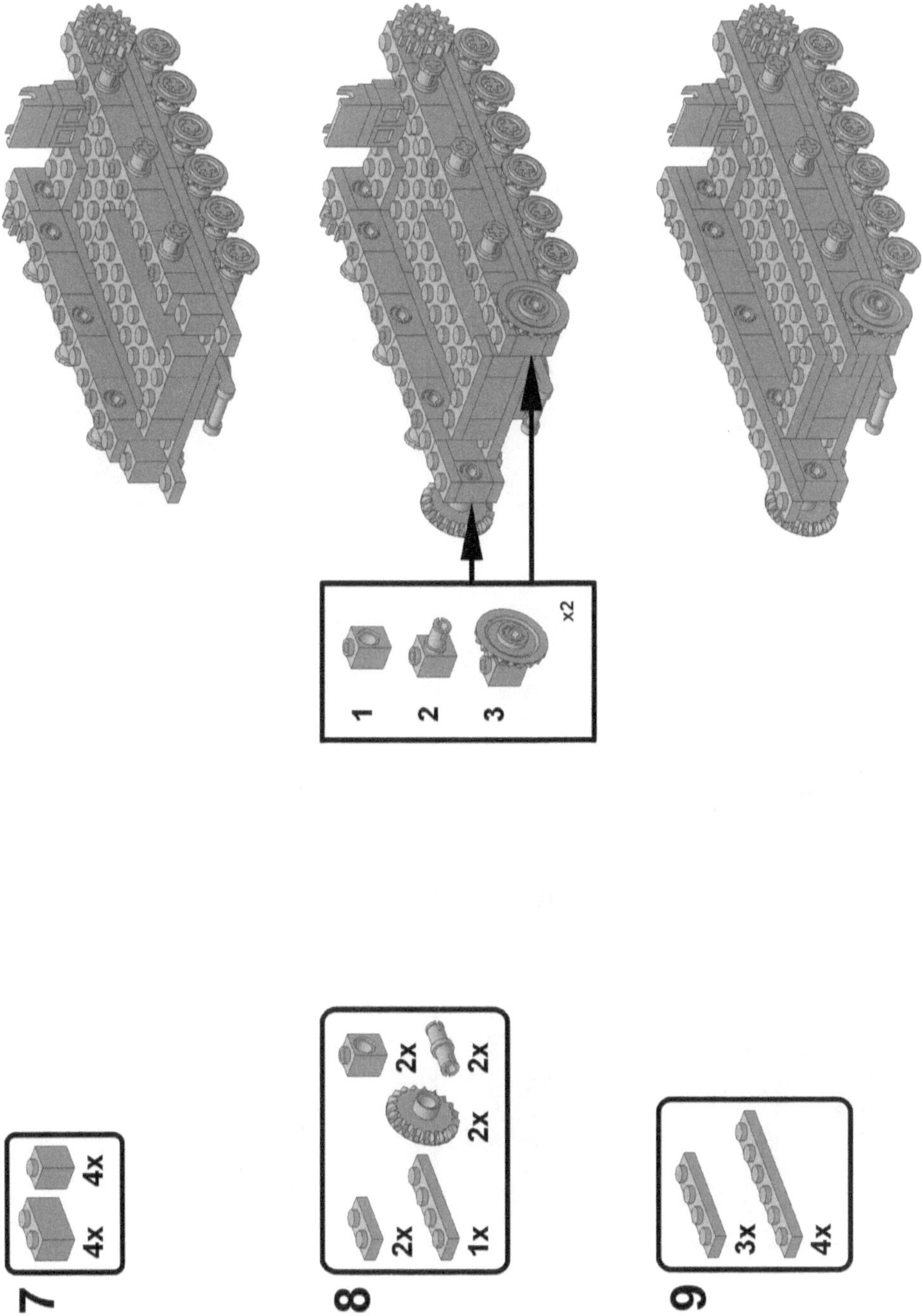

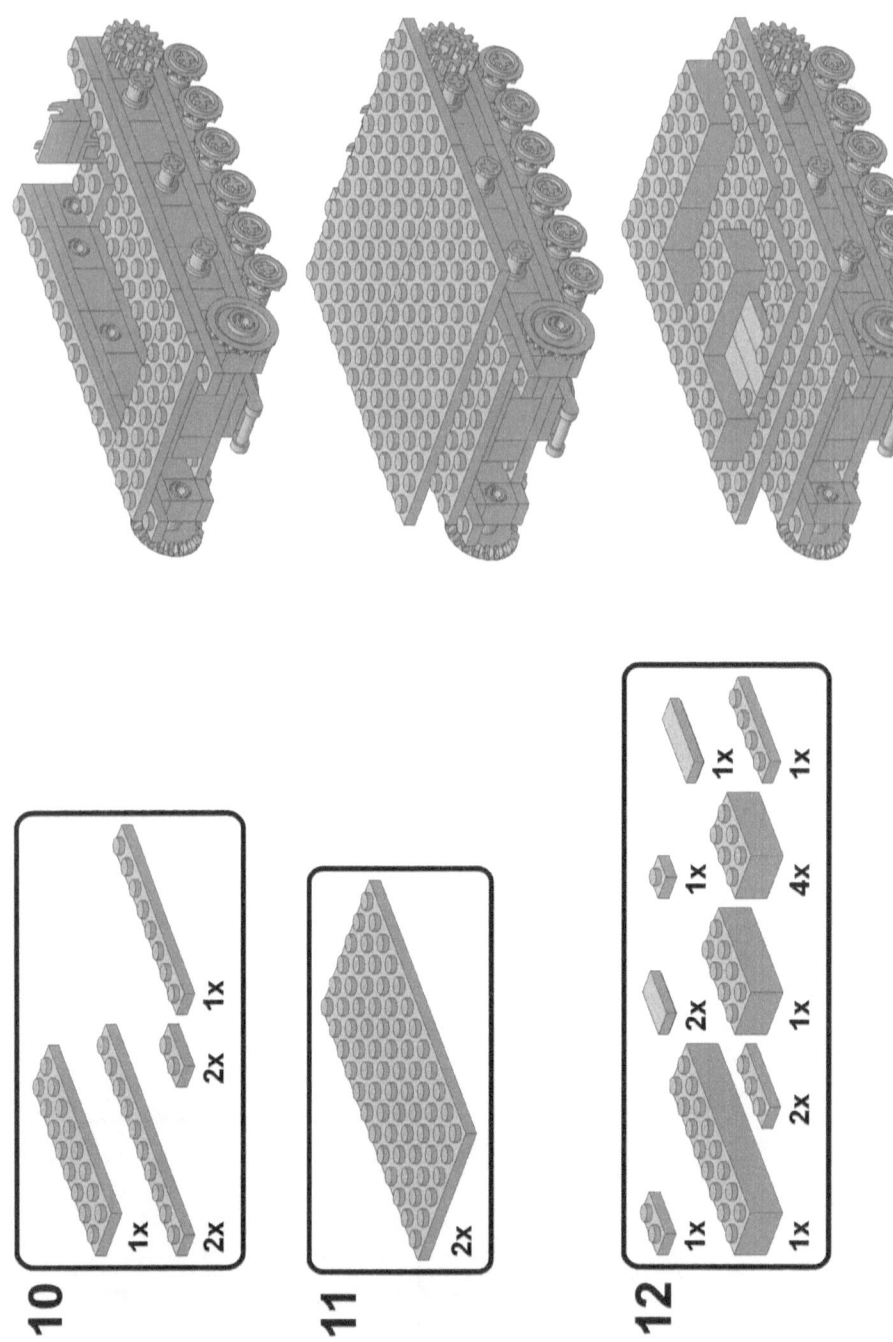

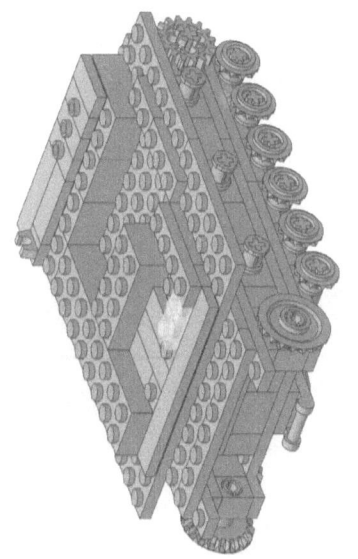
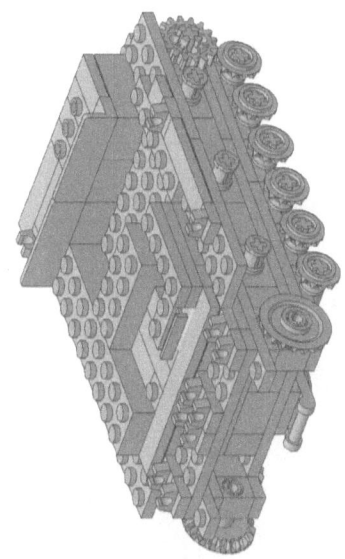
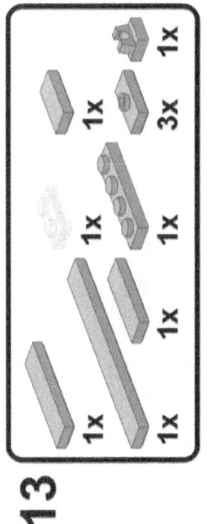
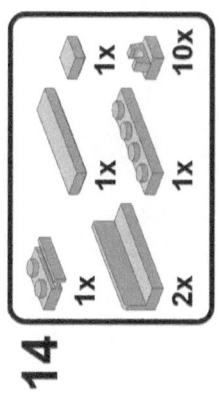

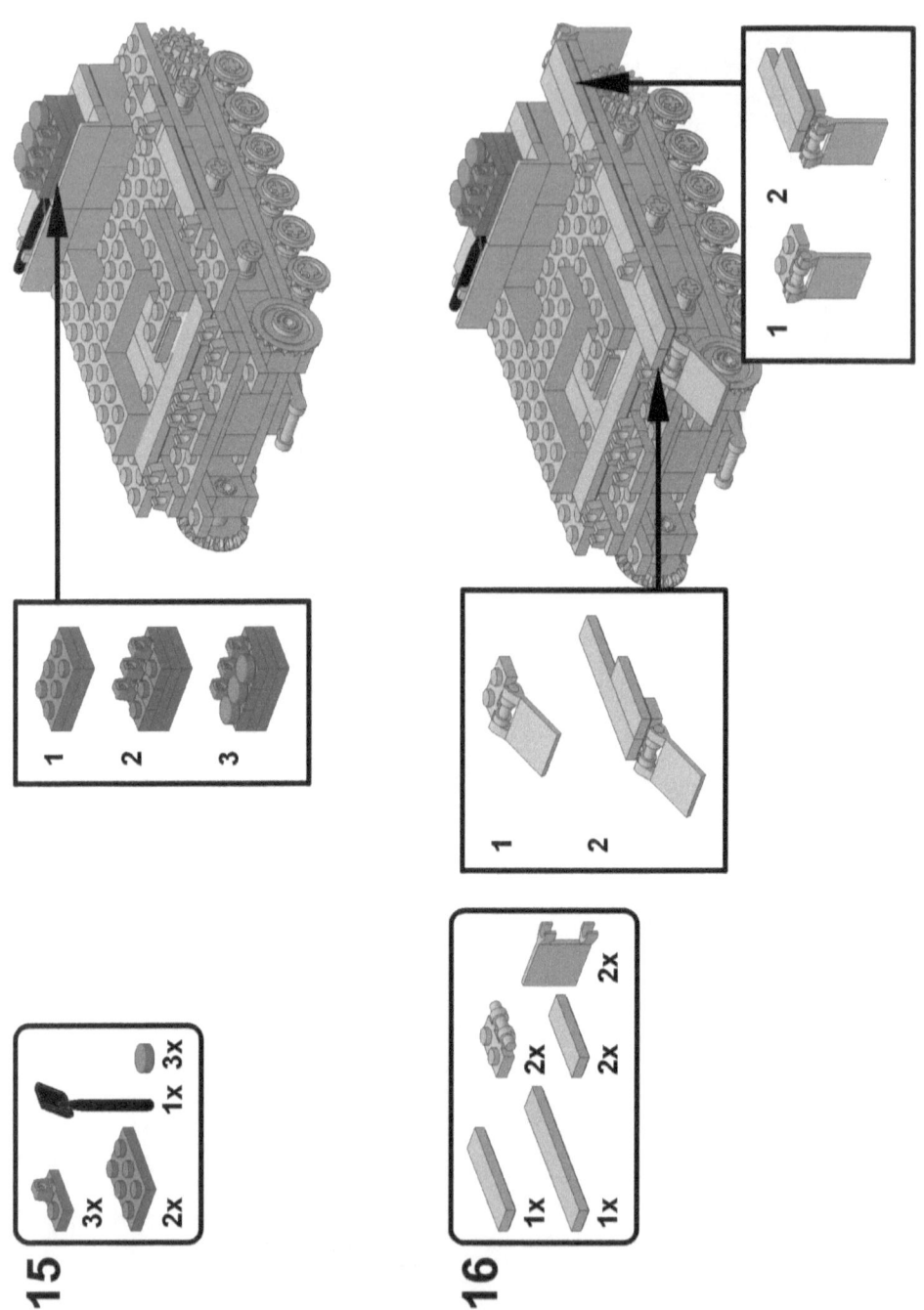

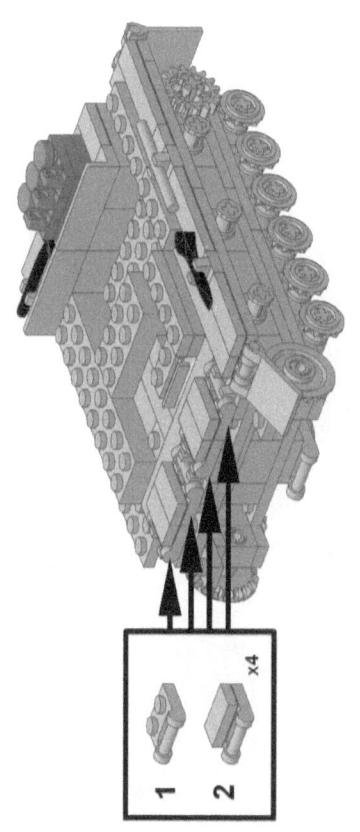
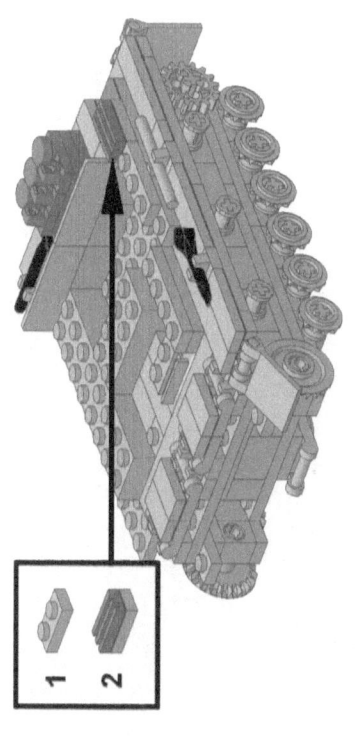
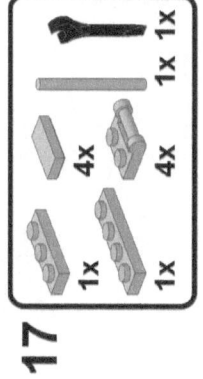
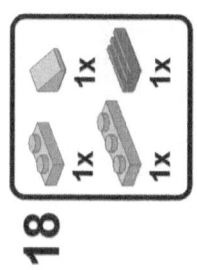

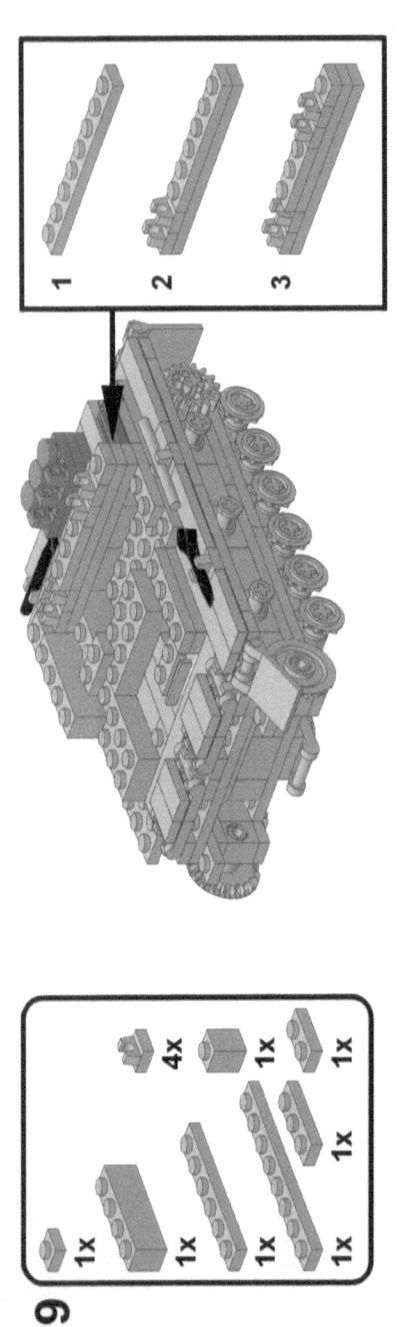
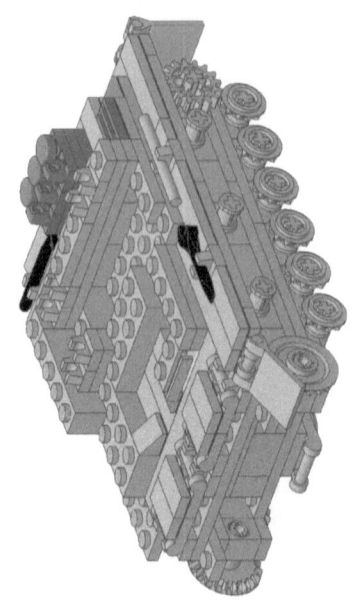
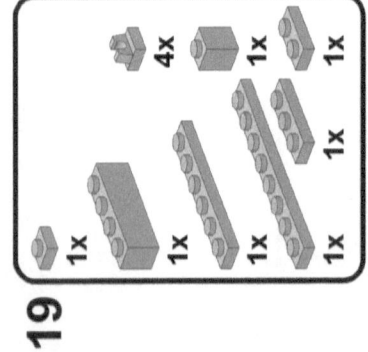
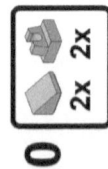

www.ww2custombrickmodels.de

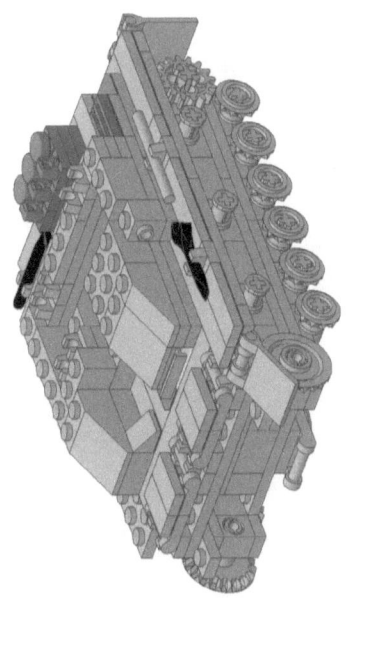
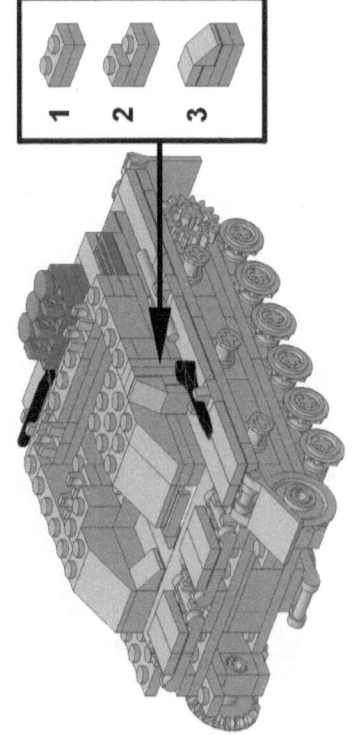
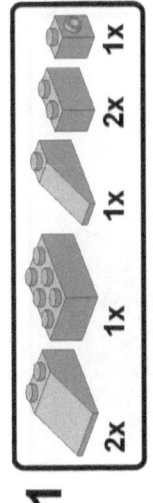
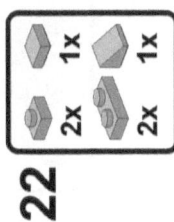

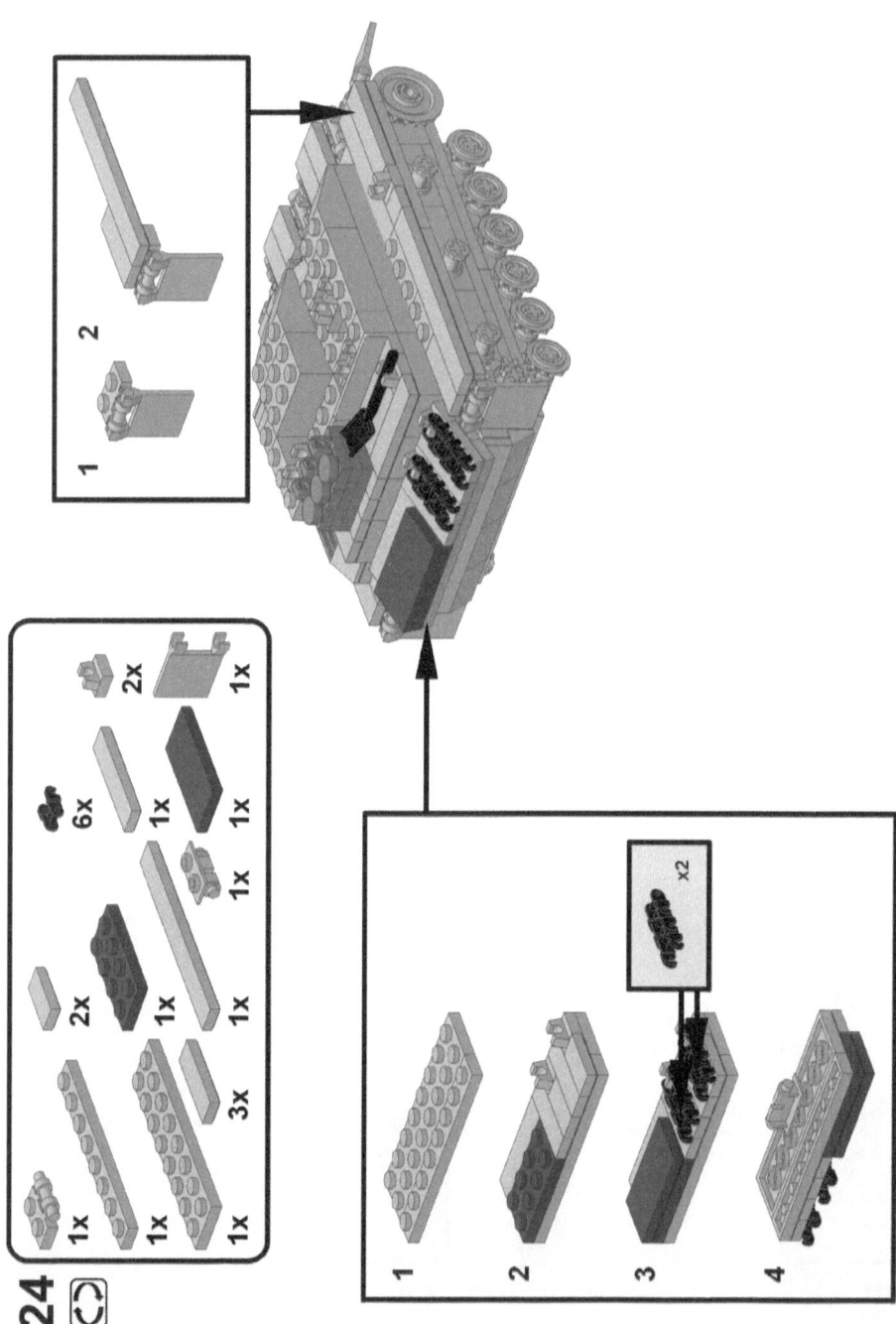

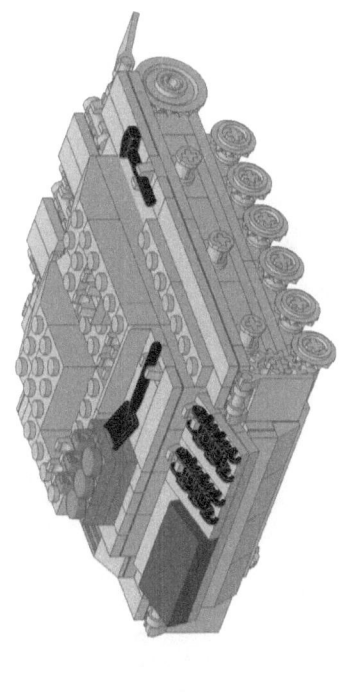
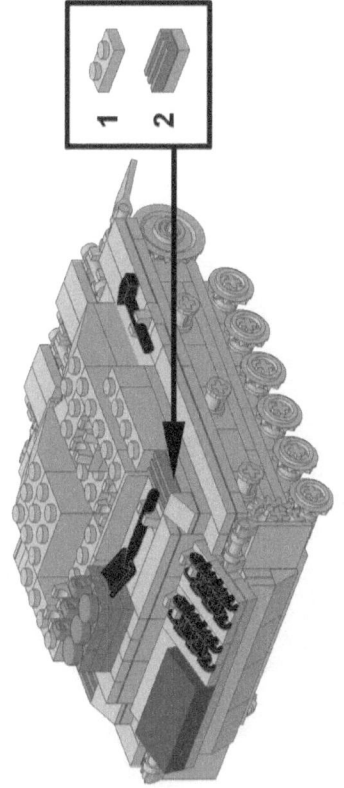
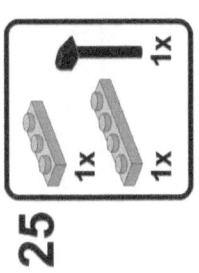

25

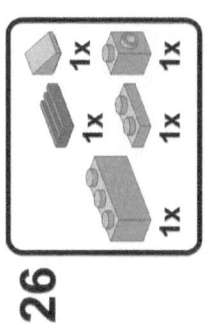

26

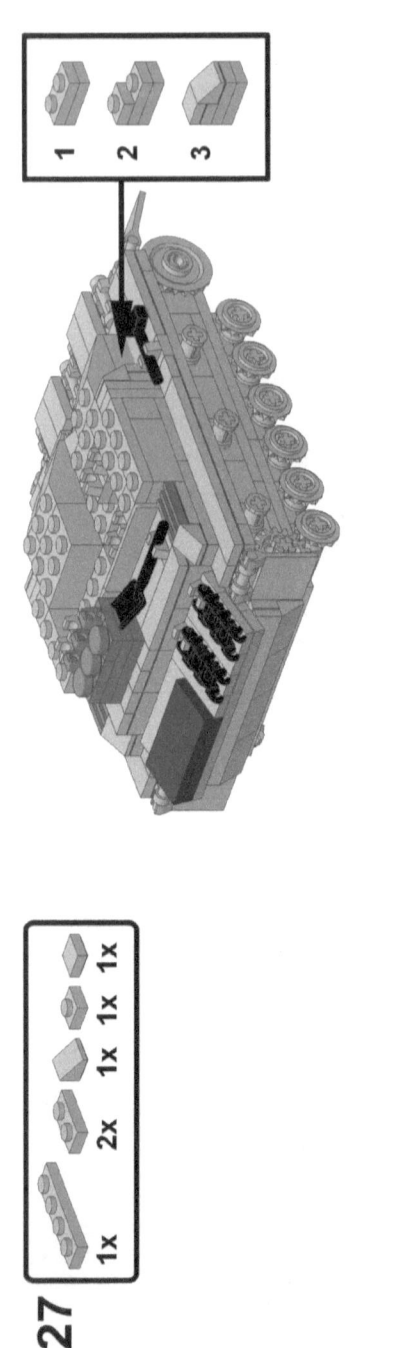
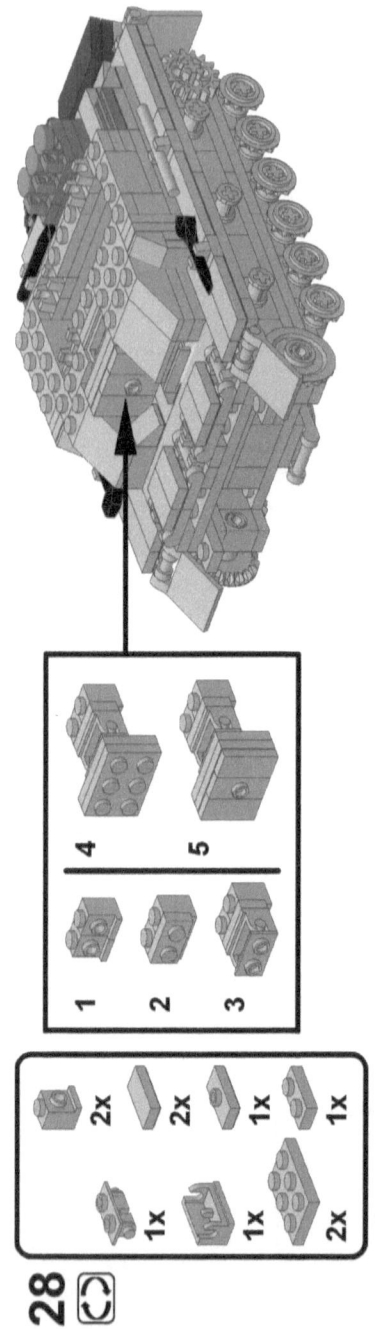

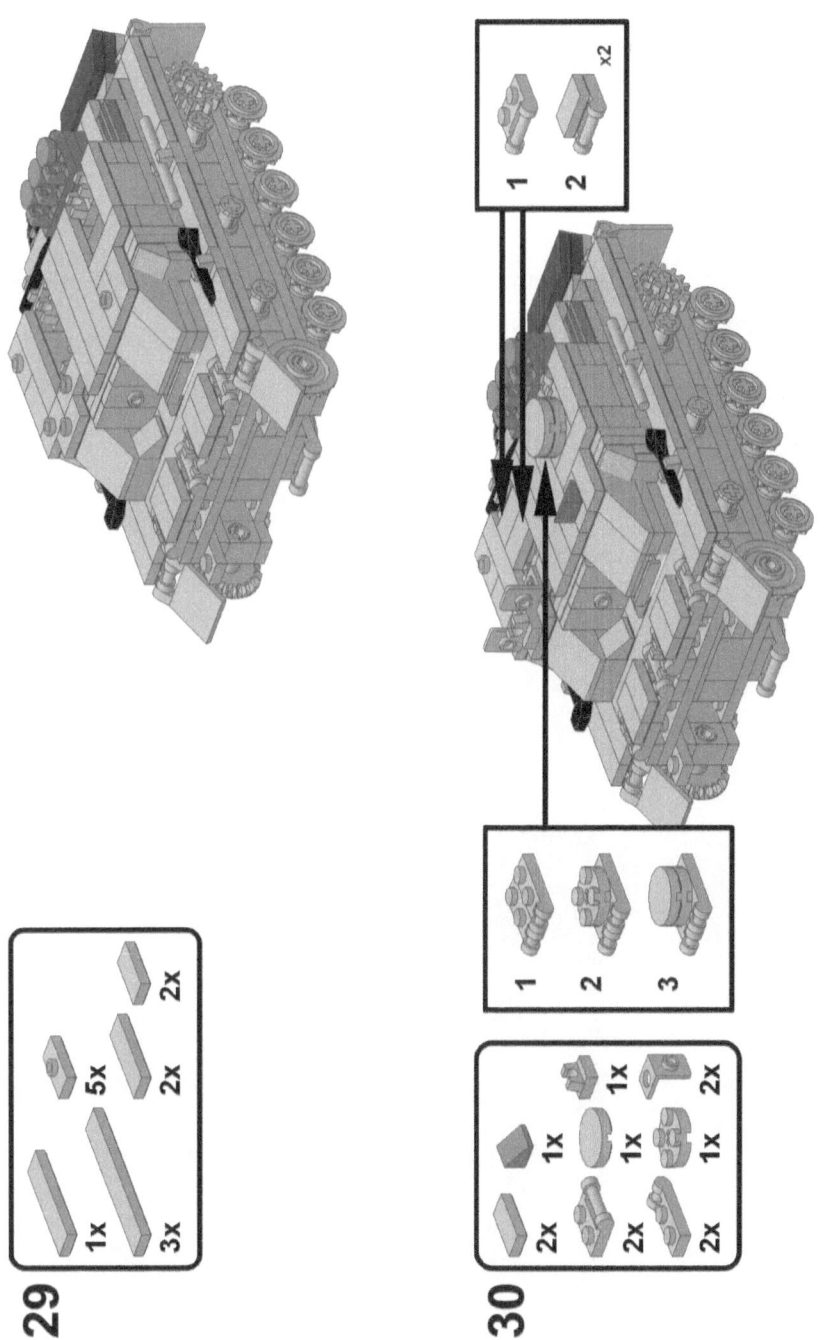

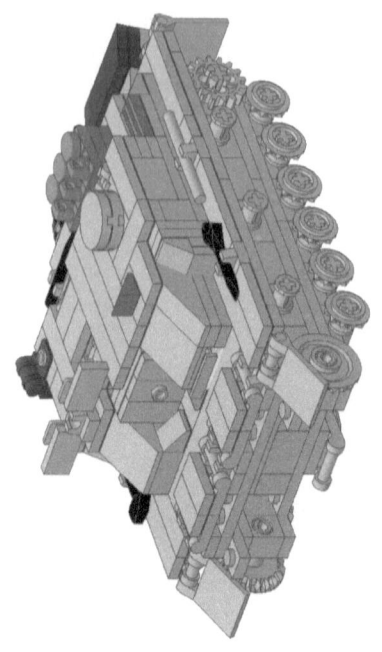

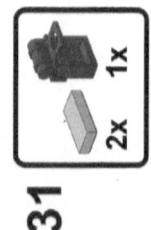

31

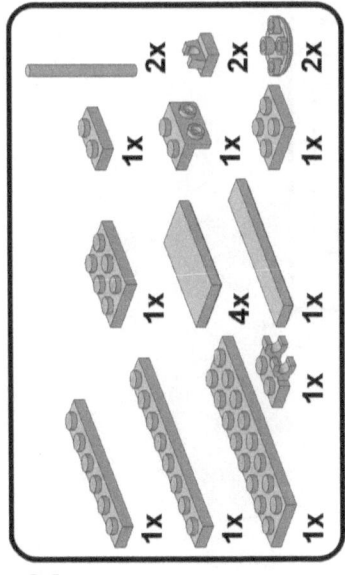

32

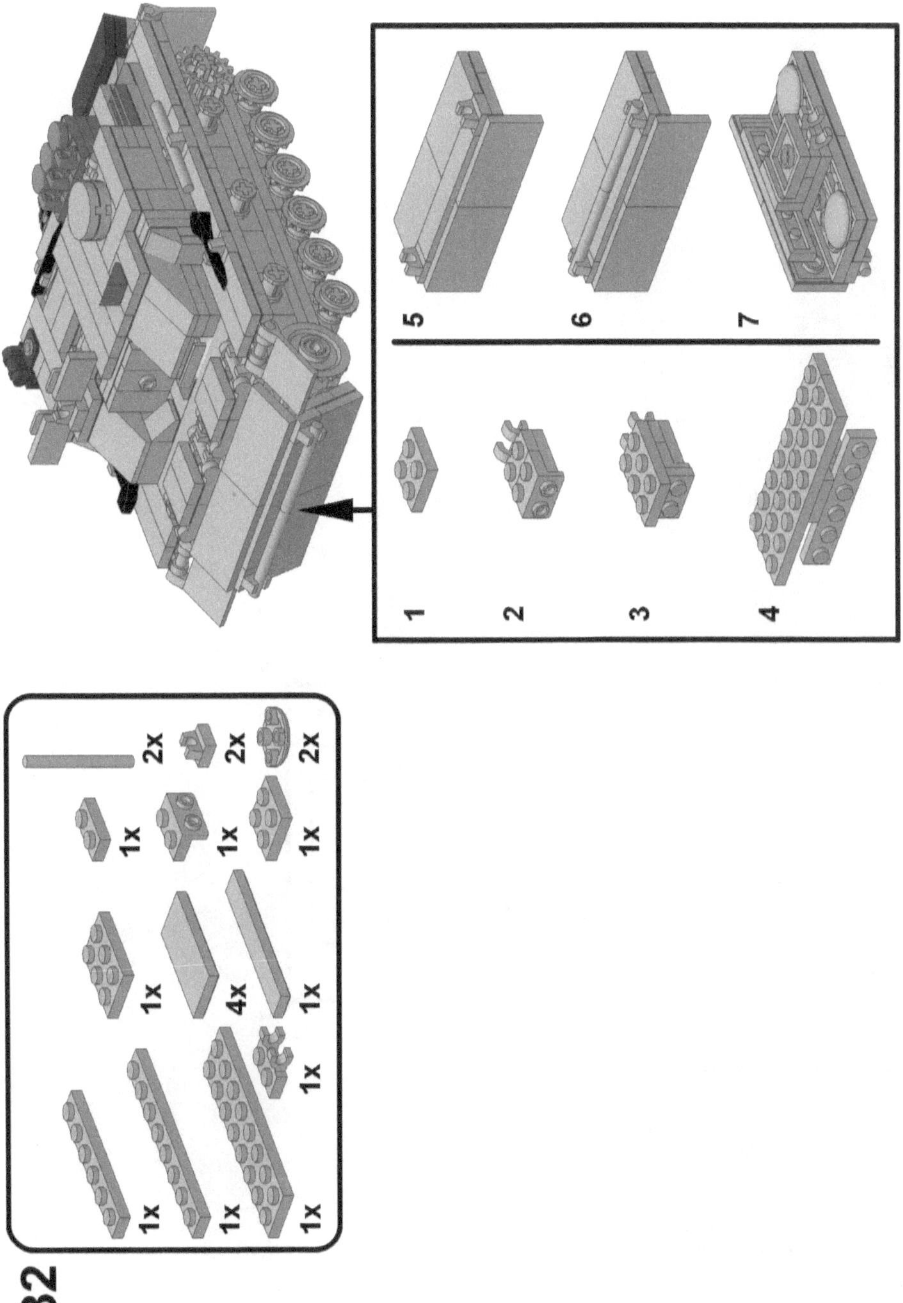

www.ww2custombrickmodels.de 97

33
1x
1x
2x
1x
2x
2x
1x
1
2
3
4
5
1
2

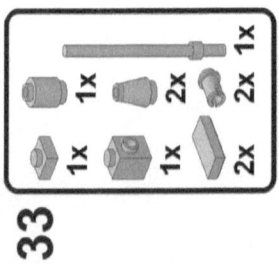

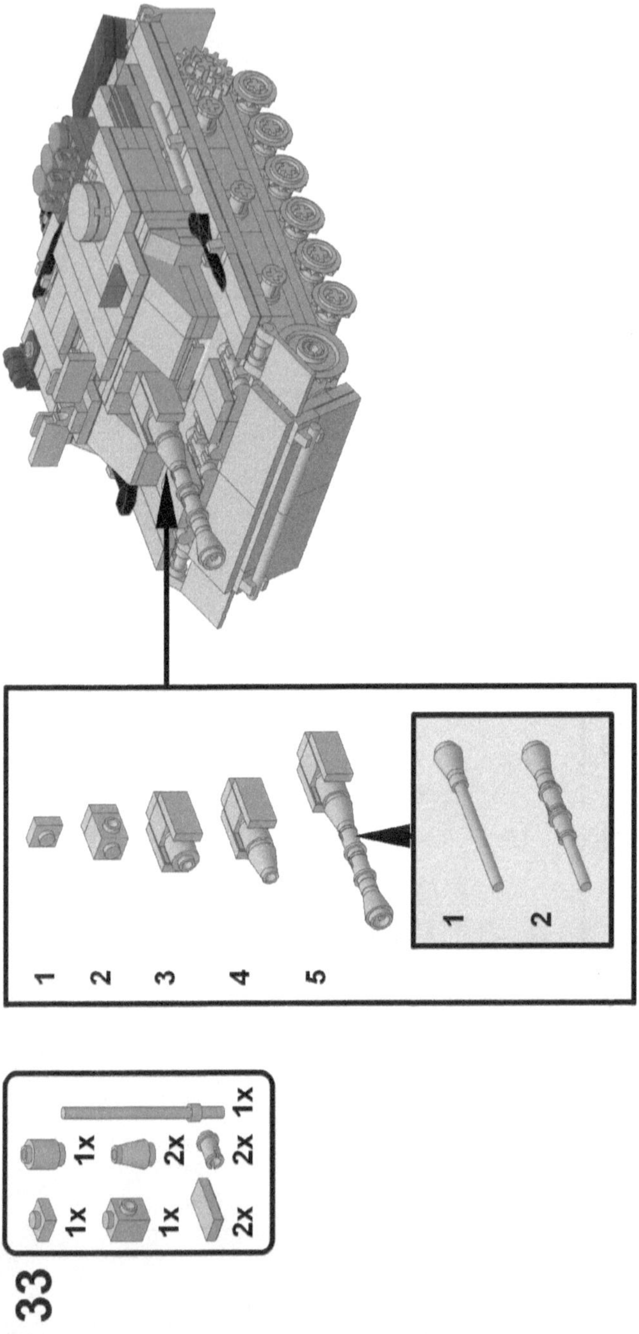

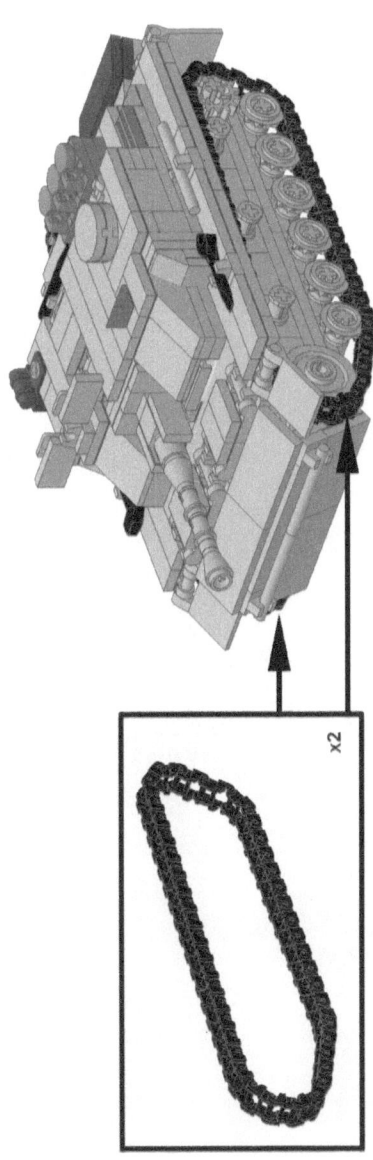

34  90x

x2

Sie sind auf der Suche nach den Teilen? Besuchen Sie einige der unten stehenden Internetseiten um mit der Hilfe der aufgelisteten Teilenummern dort das passende Teil zu finden.

Not sure where to find the parts? Try some of the websites below. You can search for the parts with the listed part numbers.

www.modernbrickwarfare.com

www.lego.de

www.bricklink.com

www.brickowl.com

Visit this website:

www.brickfactory-berlin.com/

PatSon Bricks

Custom handmade minifigures from movies and games...

Visit this website:

http://www.patsonbricks.de/

Visit this website:

https://www.facebook.com/PatSonBricks/

Visit my website:

www.ww2custombrickmodels.de

www.ingramcontent.com/pod-product-compliance
Lightning Source LLC
Chambersburg PA
CBHW031442210526
45464CB00005B/2304